T0134604

Springer Series in Adaptive Environments

The Springer Series in Adaptive Environments presents cutting-edge research around spatial constructs and systems that are specifically designed to be adaptive to their surroundings and to their inhabitants. The creation and understanding of such adaptive Environments spans the expertise of multiple disciplines, from architecture to design, from materials to urban research, from wearable technologies to robotics, from data mining to machine learning and from sociology to psychology. The focus is on the interaction between human and non-human agents, with people being both the drivers and the recipients of adaptivity embedded into environments. There is emphasis on design, from the inception to the development and to the operation of adaptive environments, while taking into account that digital technologies underpin the experimental and everyday implementations in this area.

Books in the series will be authored or edited volumes addressing a wide variety of topics related to Adaptive Environments (AEs) including:

- Interaction and inhabitation of adaptive environments
- Design to production and operation of adaptive environments
- Wearable and pervasive sensing
- Data acquisition, data mining, machine learning
- Human-robot collaborative interaction
- User interfaces for adaptive and self-learning environments
- Materials and adaptivity
- Methods for studying adaptive environments
- The history of adaptivity
- Biological and emergent buildings and cities

More information about this series at http://www.springer.com/series/15693

Nobuo Kawaguchi · Nobuhiko Nishio ·
Daniel Roggen · Sozo Inoue ·
Susanna Pirttikangas · Kristof Van Laerhoven
Editors

Human Activity Sensing

Corpus and Applications

Editors
Nobuo Kawaguchi
Institute of Innovation for Future Society
Nagoya University
Nagoya, Japan

Nobuhiko Nishio
Department of Computer Science
Ritsumeikan University
Kyoto, Japan

Daniel Roggen
University of Sussex
Brighton, UK

Sozo Inoue
Kyushu Institute of Technology
Kitakyushu, Fukuoka, Japan

Susanna Pirttikangas
Center for Ubiquitous Computing
University of Oulu
Oulu, Finland

Kristof Van Laerhoven
Ubiquitous Computing
University of Siegen
Siegen, Germany

ISSN 2522-5529 ISSN 2522-5537 (electronic)
Springer Series in Adaptive Environments
ISBN 978-3-030-13003-9 ISBN 978-3-030-13001-5 (eBook)
https://doi.org/10.1007/978-3-030-13001-5

This Springer imprint is published by the registered company Springer Nature Switzerland AG
The registered company address is: Gewerbestrasse 11, 6330 Cham, Switzerland

Preface

As part of current adaptive environments, recent technological advances have enabled the inclusion of miniature sensors (such as accelerometers or gyroscopes) on a variety of wearable and portable information devices. Most such devices utilize these sensors for simple orientation and gesture recognition only, but the recognition of more complex and subtle human behaviors from these sensors will in the future enable next-generation human-oriented computing in scenarios of high societal value (e.g., dementia care, fitness tracking, or work safety monitoring). This will require large-scale human activity corpora and much improved methods to recognize activities and the contexts in which they occur. This book deals with the challenges of designing reproducible experimental setups, running large-scale dataset collection campaigns, designing activity and context recognition methods that are robust and adaptive, and evaluating systems in the real world. Additionally, we reflect on the challenges and possible approaches to recognize situations, events, or activities outside of a statically predefined pool, which is the current state of the art, and instead adopt an open-ended view on activity and context awareness.

Following the success of five previous workshops held in conjunction with the ACM UbiComp/ISWC conferences, we have collected continuations from some of the best contributions to the past workshops as chapters in our book. This way, we share our current research on human activity corpus and their applications among the researchers and the practitioners.

We would last but not least like to thank all the contributors and workshop attendees of the past years for their valuable contributions and stimulating conversations.

Nagoya, Japan — Nobuo Kawaguchi
Shiga, Japan — Nobuhiko Nishio
Brighton, UK — Daniel Roggen
Fukuoka, Japan — Sozo Inoue
Oulu, Finland — Susanna Pirttikangas
Siegen, Germany — Kristof Van Laerhoven

Contents

Part I Modalities and Applications

1 Optimizing of the Number and Placements of Wearable IMUs for Automatic Rehabilitation Recording 3
Kohei Komukai and Ren Ohmura

2 Identifying Sensors via Statistical Analysis of Body-Worn Inertial Sensor Data 17
Philipp M. Scholl and Kristof Van Laerhoven

3 Compensation Scheme for PDR Using Component-Wise Error Models 29
Junto Nozaki, Kei Hiroi, Katsuhiko Kaji and Nobuo Kawaguchi

4 Towards the Design and Evaluation of Robust Audio-Sensing Systems 47
Akhil Mathur, Anton Isopoussu, Fahim Kawsar, Robert Smith, Nadia Berthouze and Nicholas D. Lane

5 A Wi-Fi Positioning Method Considering Radio Attenuation of Human Body 59
Shohei Harada, Kazuya Murao, Masahiro Mochizuki and Nobuhiko Nishio

Part II Data Collection and Corpus Construction

6 Drinking Gesture Recognition from Poorly Annotated Data: A Case Study 71
Mathias Ciliberto, Lin Wang, Daniel Roggen and Ruediger Zillmer

7 Understanding How Non-experts Collect and Annotate Activity Data 91
Naomi Johnson, Michael Jones, Kevin Seppi and Lawrence Thatcher

8 A Multi-media Exchange Format for Time-Series Dataset Curation 111
Philipp M. Scholl, Benjamin Völker, Bernd Becker and Kristof Van Laerhoven

9 OpenHAR: A Matlab Toolbox for Easy Access to Publicly Open Human Activity Data Sets—Introduction and Experimental Results 121
Pekka Siirtola, Heli Koskimäki and Juha Röning

10 MEASURed: Evaluating Sensor-Based Activity Recognition Scenarios by Simulating Accelerometer Measures from Motion Capture 135
Paula Lago, Shingo Takeda, Tsuyoshi Okita and Sozo Inoue

Part III SHL: An Activity Recognition Challenge

11 Benchmark Performance for the Sussex-Huawei Locomotion and Transportation Recognition Challenge 2018 153
Lin Wang, Hristijan Gjoreski, Mathias Ciliberto, Sami Mekki, Stefan Valentin and Daniel Roggen

12 Bayesian Optimization of Neural Architectures for Human Activity Recognition 171
Aomar Osmani and Massinissa Hamidi

13 Into the Wild—Avoiding Pitfalls in the Evaluation of Travel Activity Classifiers 197
Peter Widhalm, Maximilian Leodolter and Norbert Brändle

14 Effects of Activity Recognition Window Size and Time Stabilization in the SHL Recognition Challenge 213
Michael Sloma, Makan Arastuie and Kevin S. Xu

15 Winning the Sussex-Huawei Locomotion-Transportation Recognition Challenge 233
Vito Janko, Martin Gjoreski, Gašper Slapničar, Miha Mlakar, Nina Reščič, Jani Bizjak, Vid Drobnič, Matej Marinko, Nejc Mlakar, Matjaž Gams and Mitja Luštrek

Contributors

Makan Arastuie University of Toledo, Toledo, OH, USA

Bernd Becker University of Freiburg, Freiburg, Germany

Nadia Berthouze University College London, London, England

Jani Bizjak Department of Intelligent Systems, Jožef Stefan Institute, Ljubljana, Slovenia;
Jožef Stefan Postgraduate School, Ljubljana, Slovenia

Norbert Brändle Austrian Institute of Technology, Vienna, Austria

Mathias Ciliberto Wearable Technologies Laboratory, Sensor Technology Research Centre, University of Sussex, Brighton, UK

Vid Drobnič Department of Intelligent Systems, Jožef Stefan Institute, Ljubljana, Slovenia

Matjaž Gams Department of Intelligent Systems, Jožef Stefan Institute, Ljubljana, Slovenia;
Jožef Stefan Postgraduate School, Ljubljana, Slovenia

Hristijan Gjoreski Wearable Technologies Laboratory, Sensor Technology Research Centre, University of Sussex, Brighton, UK;
Faculty of Electrical Engineering and Information Technologies, Ss. Cyril and Methodius University in Skopje, Skopje, Macedonia

Martin Gjoreski Department of Intelligent Systems, Jožef Stefan Institute, Ljubljana, Slovenia;
Jožef Stefan Postgraduate School, Ljubljana, Slovenia

Massinissa Hamidi Laboratoire LIPN-UMR CNRS 7030, PRES Sorbonne Paris Cité, Villetaneuse, France

Shohei Harada Graduated School of Information Science and Engineering, Ritsumeikan University, Kyoto, Shiga, Japan

Kei Hiroi Faculty of Graduate School of Engineering, Nagoya University, Nagoya, Aichi, Japan

Sozo Inoue Kyushu Institute of Technology, Kitakyushu-shi, Fukuoka, Japan

Anton Isopoussu Nokia Bell Labs, London, England

Vito Janko Department of Intelligent Systems, Jožef Stefan Institute, Ljubljana, Slovenia;
Jožef Stefan Postgraduate School, Ljubljana, Slovenia

Naomi Johnson University of Virginia, Charlottesville, VA, USA

Michael Jones Brigham Young University, Provo, UT, USA

Katsuhiko Kaji Faculty of Information Science, Aichi Institute of Technology, Toyota, Aichi, Japan

Nobuo Kawaguchi Faculty of Graduate School of Engineering, Institutes of Innovation for Future Society, Nagoya University, Nagoya, Aichi, Japan

Fahim Kawsar Nokia Bell Labs, London, England

Kohei Komukai Graduate School of Toyohashi University of Technology, Toyohashi, Japan

Heli Koskimäki Biomimetics and Intelligent Systems Group, University of Oulu, Oulu, Finland

Paula Lago Kyushu Institute of Technology, Kitakyushu-shi, Fukuoka, Japan

Nicholas D. Lane University of Oxford, Oxford, England

Maximilian Leodolter Austrian Institute of Technology, Vienna, Austria

Mitja Luštrek Department of Intelligent Systems, Jožef Stefan Institute, Ljubljana, Slovenia;
Jožef Stefan Postgraduate School, Ljubljana, Slovenia

Matej Marinko Department of Intelligent Systems, Jožef Stefan Institute, Ljubljana, Slovenia

Akhil Mathur Nokia Bell Labs and University College London, London, England

Sami Mekki Mathematical and Algorithmic Sciences Lab, PRC, Huawei Technologies France, Boulogne-Billancourt, France

Miha Mlakar Department of Intelligent Systems, Jožef Stefan Institute, Ljubljana, Slovenia

Nejc Mlakar Department of Intelligent Systems, Jožef Stefan Institute, Ljubljana, Slovenia

Masahiro Mochizuki Research Organization of Science and Technology, Ritsumeikan University, Kyoto, Shiga, Japan

Kazuya Murao College of Information Science and Engineering, Ritsumeikan University, Kyoto, Shiga, Japan

Nobuhiko Nishio College of Information Science and Engineering, Ritsumeikan University, Kyoto, Shiga, Japan

Junto Nozaki Graduate School of Engineering, Nagoya University, Nagoya, Aichi, Japan

Ren Ohmura Toyohashi University of Technology, Toyohashi, Japan

Tsuyoshi Okita Kyushu Institute of Technology, Kitakyushu-shi, Fukuoka, Japan

Aomar Osmani Laboratoire LIPN-UMR CNRS 7030, PRES Sorbonne Paris Cité, Villetaneuse, France

Nina Reščič Department of Intelligent Systems, Jožef Stefan Institute, Ljubljana, Slovenia;
Jožef Stefan Postgraduate School, Ljubljana, Slovenia

Daniel Roggen Wearable Technologies Laboratory, Sensor Technology Research Centre, University of Sussex, Brighton, UK

Juha Röning Biomimetics and Intelligent Systems Group, University of Oulu, Oulu, Finland

Philipp M. Scholl University of Freiburg, Freiburg, Germany

Kevin Seppi Brigham Young University, Provo, UT, USA

Pekka Siirtola Biomimetics and Intelligent Systems Group, University of Oulu, Oulu, Finland

Gašper Slapničar Department of Intelligent Systems, Jožef Stefan Institute, Ljubljana, Slovenia

Michael Sloma University of Toledo, Toledo, OH, USA

Robert Smith University College London, London, England

Shingo Takeda Kyushu Institute of Technology, Kitakyushu-shi, Fukuoka, Japan

Lawrence Thatcher Brigham Young University, Provo, UT, USA

Stefan Valentin Department of Computer Science, Darmstadt University of Applied Sciences, Darmstadt, Germany

Kristof Van Laerhoven University of Siegen, Siegen, Germany

Benjamin Völker University of Freiburg, Freiburg, Germany

Lin Wang Wearable Technologies Laboratory, Sensor Technology Research Centre, University of Sussex, Brighton, UK;
Centre for Intelligent Sensing, Queen Mary University of London, London, UK

Peter Widhalm Austrian Institute of Technology, Vienna, Austria

Kevin S. Xu University of Toledo, Toledo, OH, USA

Ruediger Zillmer Unilever R&D Port Sunlight, Birkenhead, UK

Part I
Modalities and Applications

What sensors work well for detecting which activities a user is performing, and where should they place? This question has been occupying researchers for a while, and has led to several fundamental insights into the nature of activity sensor data. This first part contributes with five chapters that each exemplifies how important and characteristic the sensor modality is for activity recognition and beyond, for instance, in making localizing of a user more accurate.

A first contribution, Chap. 1, examines the optimal number of and placement of wearable sensors that are needed to suitably recognize activities in the area of rehabilitation monitoring. In a study with 16 participants, it was found that only 3 sensors were sufficient to still obtain a recognition of 83 for the F1-measure.

Following this is a chapter entitled Chap. 2, which discusses the different sensors that are typically used in wearable activity recognition research, the accelerometer, gyroscope, and magnetometer, and looks at the inherent characteristics of the data they produce. Results on several datasets with a large amount of inertial data show that statistics of inertial sensor data can be used to identify their modality.

In a third contribution, titled Chap. 3, the application of pedestrian dead reckoning is tackled through a compensation scheme. This scheme uses sparse locations and error models to correct for the typical errors that occur when persons' steps, orientation, and walking activity are monitored and used to re-create the movement trajectories of that person. In an evaluation study held at a museum, it was found that the proposed scheme improved the position evaluation metric to approximately 10 and the distance evaluation metric to approximately 7.

This is followed by a Chap. 4, in which audio inference models are explored for real-world applications. In this chapter, three empirical studies form an evaluation of the impact of hardware and environment variabilities on both cloud-scale and embedded-scale audio models. Results include the significant performance degradation of current models in the presence of ambient acoustic noise and under microphone variability.

A final chapter in this part describes the use of radio signals in Chap. 5. Here, the focus is on the localization in the presence of nearby humans that tend to influence, sometimes even block, the observed RSSI values.

Chapter 1
Optimizing of the Number and Placements of Wearable IMUs for Automatic Rehabilitation Recording

Kohei Komukai and Ren Ohmura

Abstract From the increase in the number of rehabilitation patients, precise and detailed records of rehabilitation have become difficult due to the busyness of physical therapists. This creates difficulty for quantitative analysis of rehabilitation. Therefore, we are constructing a system that automatically records rehabilitation using activity recognition techniques with wearable sensors. The system can offer the dual prospect of decreasing a practitioners' load and enabling quantitative analysis of rehabilitation. In general, a large number of wearable sensors are demanded to be placed on each part of a patient's body for accurate activity recognition. However, managing a larger number of wearable sensors incurs significant time and effort of therapists and patients to apply them and adds to a patients' discomfort, consequently it significantly decreases practicality. Therefore, we investigate the suitable number and positions of wearable sensors for activity recognition for rehabilitation. Experiments were carried out with 16 healthy subjects wearing seven wearable inertial measurement units (IMUs). The subjects performed 10 different rehabilitation activities chosen by a qualified physical therapist as typical ones used in real rehabilitation therapy. The activities were recognized while reducing the number of sensors with all combinations of sensor placements using six classification algorithms. Then, the accuracy on each setting was examined. As a result, 0.833 of F-measure value was obtained when using three sensors on the waist, right thigh, and right lower leg.

1.1 Introduction

In recent years, with the progression of an aging society, the number of patients requiring rehabilitation treatment is increasing for the purpose of recovery of motor function and improvement of Activities of Daily Living (ADL). Generally, a prac-

K. Komukai
Graduate School of Toyohashi University of Technology, Toyohashi, Japan

R. Ohmura (✉)
Toyohashi University of Technology, Toyohashi, Japan
e-mail: ren@tut.jp

N. Kawaguchi et al. (eds.), *Human Activity Sensing*,
Springer Series in Adaptive Environments,
https://doi.org/10.1007/978-3-030-13001-5_1

titioner, called a physical therapist (PT), treats many patients in a day, and taking precise records of each patient is very difficult due to the high workload. Instead, they record treatment given to each patient by hand on some form of notepaper. Accordingly, formal rehabilitation records for all patients are recorded through a combination of written notes and the therapist's memory after completing the day's therapy. Thus, rehabilitation records can be imprecise. Moreover, conducting a quantitative assessment of rehabilitation becomes very difficult for this reason (González-Villanueva et al. 2013; Mori et al. 2015; Vincent et al. 2008).

In order to solve this problem, we are developing an automatic rehabilitation recording system using an activity recognition technique with wearable sensors, aiming to acquire accurate and detailed rehabilitation records. The system can enable quantitative evaluation and PT burden reduction. There are existing researches for rehabilitation recognition using image sensors, such as RGB-D sensors. However, the placement of the image sensor and a patients' privacy are problematic when using such methods. Instead, this study uses wearable sensors for activity recognition to address these problems.

In general, in order to recognize a number of complex actions accurately, a user is required to wear a large number of sensors. However, this demands a significant effort and time for both the physical therapist and the patient through the course of putting them on at the beginning of therapy and taking them off once finished. This creates a significant problem in a real physical therapist environment. Therefore, in order to increase practicality, it is necessary to balance the number of sensors worn by the patient with the required accuracy.

Therefore, in this research, the optimum number and placement of sensors are investigated in order to enable an automatic rehabilitation recording system. First, seven inertial measurement units (IMUs) are attached to each part of a subjects body and collect data during rehabilitation activities, which are commonly administered as real rehabilitation treatments and were chosen by a real physical therapist. Based on the obtained data, while the number of sensors is reduced and the combinations of sensor placements are changed, the recognition accuracy is evaluated during which the classification algorithms were changed. Then, the optimal balance of the number, positions, and accuracy was examined from the obtained results. In addition, a suitable classification was considered for the sensor setting. The results of the preliminary experiment are shown in Komukai and Ohmura (2018). From Komukai and Ohmura (2018), the parameters in classification algorithms are carefully adjusted, and the results of each experiment's settings are re-evaluated, especially for the DNN model.

1.2 Related Work

Studies on activity recognition using an RGB-D camera have been conducted to evaluate cognitive rehabilitation (Minamoto et al. 2016; Ikegaya et al. 2016). In these studies, one or more cameras are necessary to be installed in each place where therapy is performed. However, the installation of the image sensor requires a significant effort and the recognition is enabled only in those the places. In addition, some patients have genuine privacy concerns. Rehabilitation recognition using wearable sensors can solve these dependencies on locations and address privacy issues.

There is a study that recognizes rehabilitation using wearable sensors. (González-Villanueva et al. 2013). However, the study recognizes only postures, such as standing and sitting. and doesn't include motions, such as swinging a shoulder, while many rehabilitation include them. There are studies focusing on recognition accuracy along the number of wearable sensors. In Gao et al. (2014), subjects equipped four tri-axis accelerometers on the subject's left lower arm, lower back, and thigh, and accuracy on activity recognition between single and multiple wearable sensors was compared. The results showed that recognition accuracy with four sensors was better than using only one sensor. However, this research does not investigate the placement and combination of wearable sensors, which is considered in our research.

Some studies focused on the optimal placement of wearable sensors. A study (Pannurat et al. 2017), evaluated the appropriate placement of wearable sensors, as well as classification algorithms and feature values. They found that the activity recognition accuracy reached 96% with sensors worn on the front of the waist, side of the waist, chest, and thigh. However, the experiment was conducted with static activities, namely 'postures', except for walking, such as sitting, standing, and sleeping. Most of the rehabilitation activities evaluated in this study consist of the dynamic motions of the patient's body, such as standing up and the movement of shoulders.

1.3 System Configuration for Experiment

1.3.1 System Overview

For evaluating activity recognition accuracy affected by the number of sensors and their placements in a rehabilitation therapy, we developed a system that collected sensor data from a patient's body with multiple sensors. The overview of the system we implemented for our experiment is shown in Fig. 1.1. This system consists of seven wearable sensors (TSND121 by ATR-promotions, Fig. 1.2, http://www.atr-p.com/products/TSND121.html), an Android terminal (MediaPadT2 8Pro, MSM

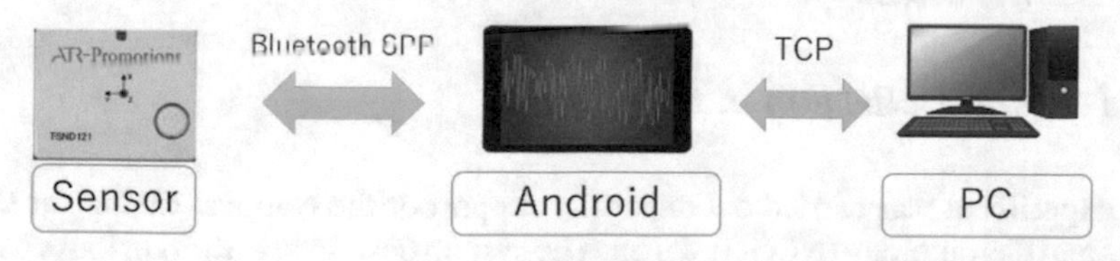

Fig. 1.1 System overview

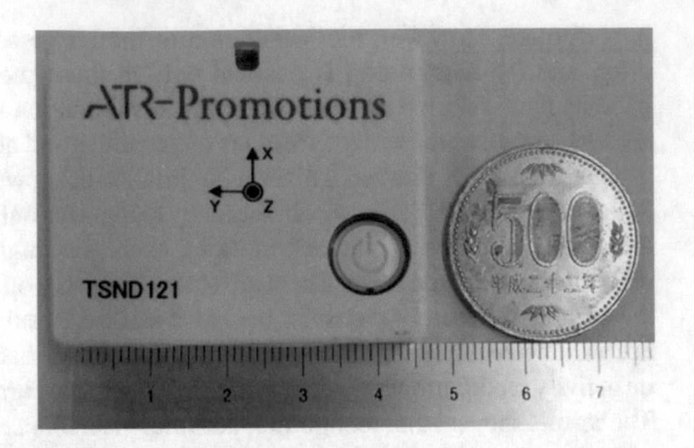

Fig. 1.2 Wearable sensor

8939, RAM 2G, Android 6.0 by HUAWEI), a PC (Windows10, i5-5330U, RAM4G, E5250 made by DELL). The sensor connects to the Android terminal via Bluetooth SPP communication. Data obtained by the sensors are transmitted to the Android terminal at first, and then the data is transferred to PC by Wi-Fi communication. On the PC, the integrity of the data is checked.

The Android terminal has the function to control sensors, such as starting and stopping the recording of sensor data and changing the sampling setting (e.g. sampling rate) of sensors. We assume that a PT holds the Android terminal and controls the sensors while a patient wears the sensors in a practical environment.

1.3.2 Wearable Sensor

In our research, a patient wears wearable sensors on their body. The specification of the wearable sensor embeds a tri-axial accelerometer and a gyro, which are sampled on 1 kHz at the maximum. The range of acceleration data is ±16 G, and the range of angular velocity is ±2000 dps. While the sensor data is transmitted via Bluetooth communication, the data can also be stored in the sensor simultaneously. Even when the Bluetooth connection is lost, the data can be recovered with the data stored in the sensor.

1.4 Experiment

1.4.1 Data Collection in Experiment

Our experiment was carried out under the support of the National Center for Geriatrics and Gerontology (NCGG), Japan. The system described in the previous chapter was used for the experiment. Subjects participating in this experiment were healthy

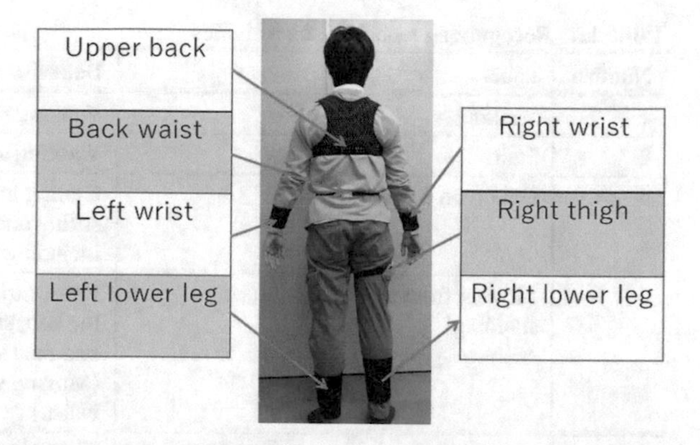

Fig. 1.3 Positions of wearable sensors

subjects, such as real practitioners, staff of NCGG, and students from our own university. The wearable sensors were attached to seven parts of the subject's body as shown in Fig. 1.3. In the experiments, sensor data were sampled with 100 Hz sampling rate with the range of ± 16 G for the accelerometer and ± 2000 dps for the gyro. From the experiments, a total of 762,422 sensor data were collected from 16 subjects.

Mori et al. devised the Basic Movement Scale (BMS), an index for measuring Basic Movement. (Mori et al. 2015). With advice from NCGG considering rehabilitation motions and BMS index, 10 kinds of rehabilitation motions, which are actually and frequently applied to real patients and PTs consider important, were decided as the target of activity recognition (Table 1.1).

1.4.2 Feature Values

From the obtained sensor data, feature values for activity recognition are calculated from data in a window of the sliding window technique, which is applied with the setting of 256 samples as its window size and 50% shift. The extracted feature values were average, variance, maximum value, minimum value, peak of frequency, energy, between sensors correlation coefficient of each axis for both accelerometer data and gyro data. The average was calculated by Eq. (1.1), the variance was calculated by Eq. (1.2), the maximum value was calculated by Eq. (1.3) and the minimum value was calculated by Eq. (1.4). After FFT was applied, peak frequency and energy were calculated by Eqs. (1.5) and (1.6), respectively. Here, x_i denotes a sensor data sample, N denotes the number of data samples in a window, f_i represents the i the frequency component by FFT. The correlation coefficient was calculated by Eq. (1.7), where S_x is standard deviation of axis of x and S_y is standard deviation of axis of y, and S_{xy} is covariance of axis of x and y. The correlation coefficients are calculated between axes only in a single sensor.

Table 1.1 Recognizing rehabilitation activities

Number	Label	Behavior details
1	Walking	Walking round on flat ground
2	Stairs	Walking and descending stairs
3	Get up on a bed	Getting up from laying on a bed and sitting on the bed (Moving whole body including arms and waist.)
4	Moving from a bed to chair (with standing)	From sitting on a bed, standing up from the bed, standing up completely once, and then sitting on a chair near the bed (Moving whole body including arms and waist.)
5	Moving from a bed to chair (without standing)	From sitting on a bed, standing up from the bed, and sitting on a chair near the bed without completely standing up (Moving mainly arms and knees a little.)
6	Standing and Sitting	Standing up and sitting down repeatedly (Moving mainly knees arms a little.)
7	Balance on sitting	While sitting, tilt the body to the left and right repeatedly (with a thought to make the base of support (BOS) small as much as possible) (Spreading both arms.)
8	Balance on standing balance	While standing, tilt the body to the left and right repeatedly (with a thought to make the base of support (BOS) small as much as possible). Both hands spread to the left and right
9	Moving range of the shoulder joint	Moving arm with assistance while lying on a bed (for increasing the moving range of shoulder joint) (PT moves arms, not by subjects themselves.)
10	Moving range of the hip joint	Moving leg with assistance while lying on a bed (for increasing the moving range of hip joint) (PT moves legs, not by subjects themselves.)

$$\bar{x} = \frac{1}{N}\sum_{i=1}^{N} x_i \tag{1.1}$$

$$s^2 = \frac{1}{N}\sum_{i=1}^{N} (x_i - \bar{x})^2 \tag{1.2}$$

$$M = \max_{i \in N}(x_i) \tag{1.3}$$

$$m = \min_{i \in N}(x_i) \tag{1.4}$$

Table 1.2 Instances of subjects

User ID	Instances	User ID	Instances
0	214	8	131
1	306	9	145
2	939	10	176
3	937	11	161
4	812	12	160
5	944	13	195
6	908	14	211
7	133	15	204
Total		6576	

$$G_{peakF} = \frac{2\pi \cdot argmax_i |f_i|}{T} \tag{1.5}$$

$$G_{energy} = \frac{1}{T} \sum_{i=1}^{N} |f_i|^2 \tag{1.6}$$

$$r = \frac{S_{xy}}{S_x \times S_y} \tag{1.7}$$

1.4.3 The Number of Data Instances

From the obtained data, feature values are extracted as mentioned above, and 6576 instances are obtained. Table 1.2 shows the number of instances of each subject. Although the number of instances is unbalanced, all data are used for training and testing.

1.4.4 Classifier

In this experiment, Support Vector Machine (SVM), Naive Bayes (NB), Nearest Neighbor (1NN), k-Nearest Neighbor (k = 3, 3NN), Random Forest (RF), and Deep Neural Network (DNN) were examined as the variation of the classification algorithms. The feature values described above were used for classification algorithms except DNN. For these algorithms, sklearn is used for their implementation. DNN input is window data. This window data is the same as the data used to make Eqs. (1.1)–(1.7) feature data.

First, we describe the parameters of SVM, NB, 1NN, 3NN, RF classifier. NB used GaussianNB algorithms. SVM used Linear SVM. For each parameter, the loss function is the square of the hinge loss, the multi-class strategy is n_classes one-vs-rest classifiers, and the maximum iteration number is 1000 times. 1NN, 3NN used the BallTree algorithm. For each parameter, leaf size is 30, metric to use for distance computation is an Euclidean metric. For the parameters of RF, the number of trees is 10 in forest, the maximum depth is not limited, and the minimum number of samples necessary for division is two.

Next, the parameters of DNN will be described. DNN was implemented using Convolutional Neural Network (CNN). The DNN model is shown in Fig. 1.4. DNN consists of two convolution layers, two Pooling layers and an output layer. As the characteristic of NN, the number of input dimensions must be constant. However, in this research, the number of sensors used change depending on the combination. It is possible to obtain 6-axis data from 1 sensor (3-axis accelerometer, 3-axis angular velocity), the number of dimensions of the sensor data (N) can be determined by formula (1.8). Therefore, the number of input dimensions (X) can be obtained by Eq. (1.9). At this time, in order to uniform to the classifiers other than DNN, the DNN input data is window size = 256. (For example, when six sensors are used, $N = 36$, so $X = (36, 256)$.) When learning DNN, the loss function used is binary cross entropy, the activation function of output layer used is "softmax", the convolution layer used is "ReLU". Also, the optimizer is RMSprop, the stride of the convolution layer is 2, the pooling algorithm uses Max Pooling.

$$N = used\ sensor\ num \times 6 \tag{1.8}$$

$$X = (N, \text{window size}) \tag{1.9}$$

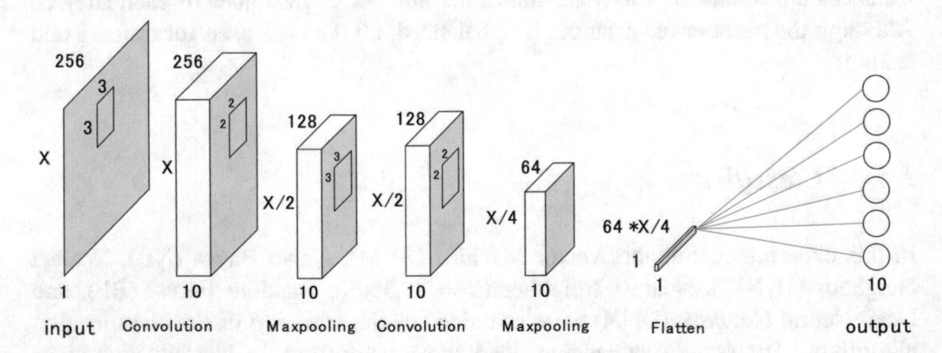

Fig. 1.4 DNN model

1.4.5 Evaluation Method

For evaluating the accuracy of each condition, the Leave-one-subject-out-cross validation method was used. As the parameter of recognition performance, F-measure value averaged on subjects was calculated.

1.5 Results and Discussion

1.5.1 Classifier

As the first evaluation, we evaluated a suitable classification algorithm without consideration of the combinations of sensor positions. The Best (Averaged F-measure: 0.796), the 2nd best (Averaged F-measure: 0.778), and the 3rd best (Averaged F-measure: 0.765) F-measure values of classifiers along the number of sensors are shown in Figs. 1.5, 1.6, and 1.7, respectively. These figures show that RandomForest gives the best performance with more than 0.1 higher than the other classifiers in all conditions of the number of sensors. Therefore, RandomForest is suitable for recognizing rehabilitation activities. From the next examination, RandomForest is used as the classifier.

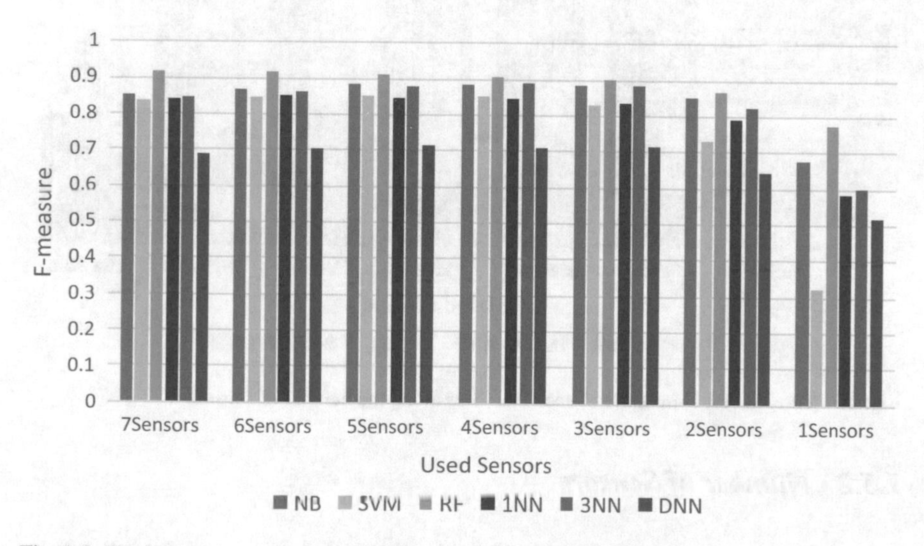

Fig. 1.5 The best F-measure of each classifier along the number of sensors

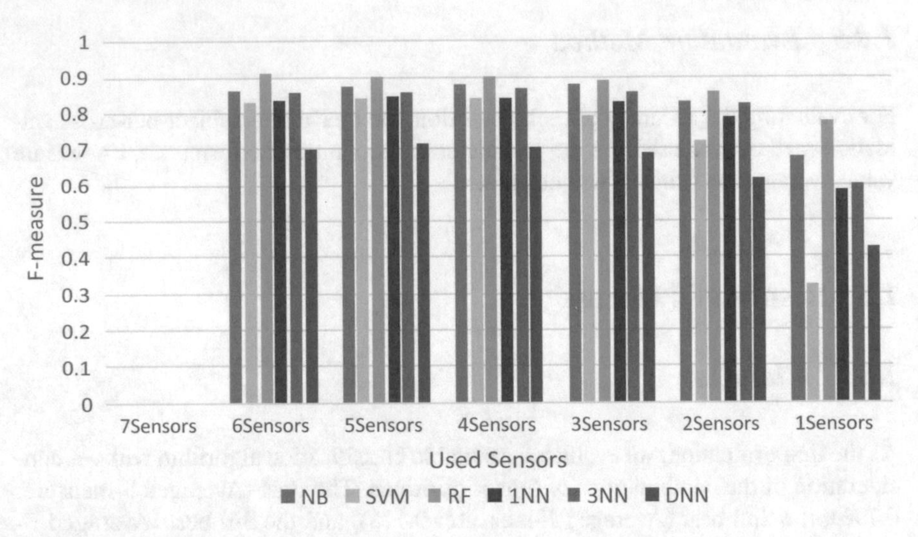

Fig. 1.6 The 2nd best F-measure of each classifier along the number of sensors

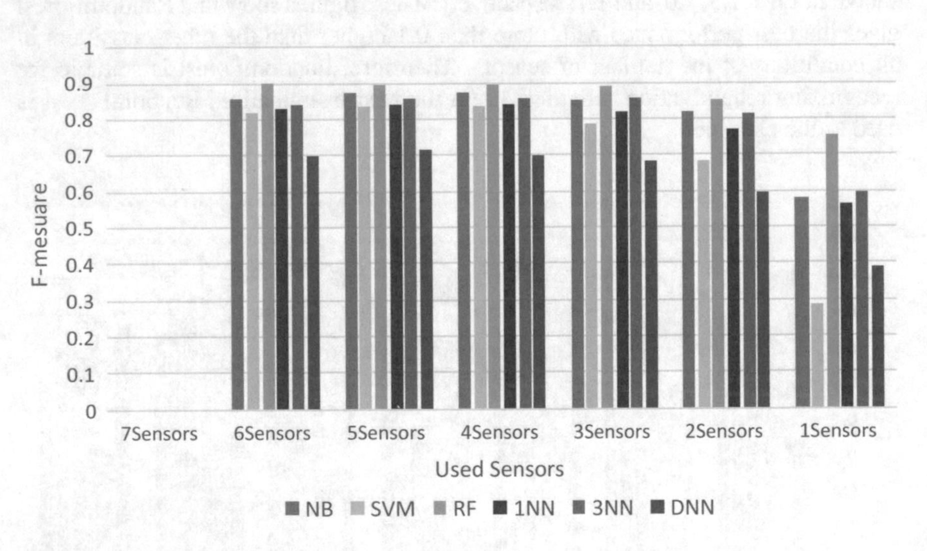

Fig. 1.7 The 3rd best F-measure of each classifier along the number of sensors

1.5.2 Number of Sensors

From the result of the previous evaluation, RandomForest is used for the followings, and then, the suitable number of wearable sensors was evaluated. Figure 1.8 shows the upper three sets of F-measure values along the number of sensors when using RandomForest. The F-measure value gets 0.915 when using seven sensors, and 0.902 when using three sensors. From these results, the performance gap between seven

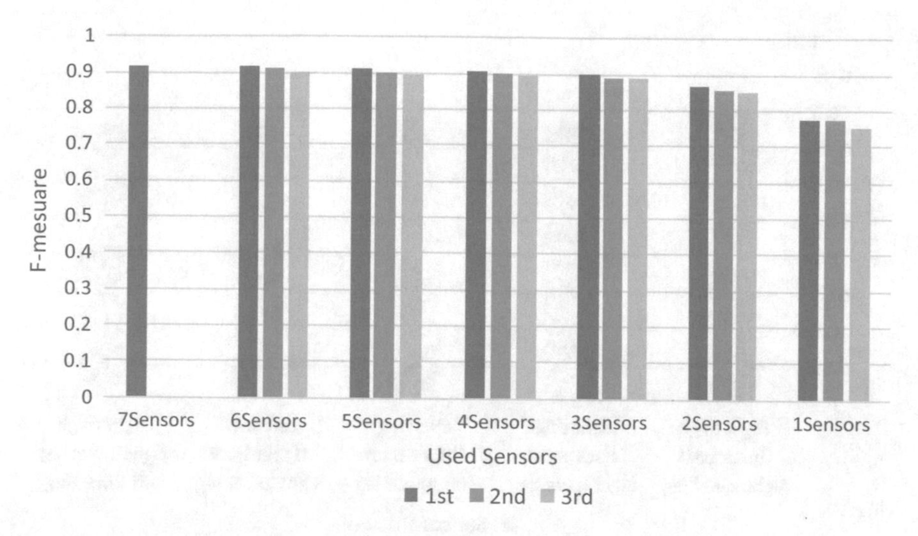

Fig. 1.8 The top three F-measure values along the number of sensors with RF

sensors and three sensors is only 0.01. Meanwhile, the F-measure on two sensors goes down to 0.866, which is decreased from three sensors by 0.036. Moreover, from opinions given by NCGG staff, three sensors are acceptable for actual use. Therefore, we conclude that using three sensors is suitable for recognizing rehabilitation motions.

1.5.3 Combination of Sensor Positions

Finally, a suitable combination of sensor positions was examined. Figure 1.9 shows the top five F-measure values with three sensors and RandomForest. From this result, the sensor positions of the best F-measure value are "center of back (upper back), right thigh, and right lower leg". However, it was found that there was a difficulty to equip a sensor on the patient's upper back due to the variation of patients' physique and gender during the experiment. Thus, the upper back is not a desirable place to attach a wearable sensor in practical usage. Instead, the combination of "right thigh, back waist, and right lower leg" gives the highest F-measure value without the upper back. In addition, the difference between F-measure value of "upper back, right thigh, right lower leg" and "right thigh, back waist, right lower leg" is quite short, 0.01. Thus, we conclude the combination of "right thigh, back waist, and right lower leg" as the suitable sensor positions.

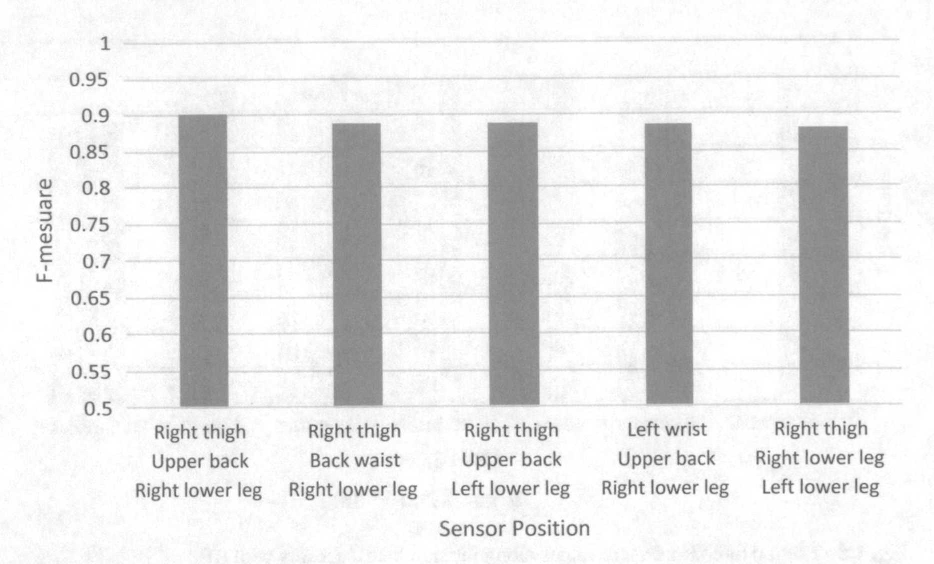

Fig. 1.9 The top five F-measure values on combinations of sensor positions using three sensors and random forest

1.6 Conclusion

With the aim of achieving a practical system to automatically record rehabilitation therapy with wearable sensors, this study explored the optimal number and positions. Then, we evaluated recognition accuracy along the number and combinations of the placement of sensors, as well as a suitable classification algorithm.

We developed a system for recording rehabilitation motions with wearable IMU sensors, in which the wearable sensor data is transferred to an Android terminal by Bluetooth communication and then transferred to a PC by Wi-Fi communication. The system minimizes the practitioners' recording effort of rehabilitation motions and allows using any environment where rehabilitation is performed.

In our experiment, data was collected from 16 subjects. The subjects wore seven wearable sensors, which collect tri-axial acceleration and angular velocities, on seven points on the subjects' body. The data were sampled at 100 Hz while subjects performed 10 kinds of rehabilitation motions suggested by a qualified rehabilitation practitioner.

From the results of the experiments, three sensors on "right thigh, back waist, and right lower leg" with RandomForest achieved 0.833 of F-measure value, while seven sensors give 0.847 and three sensors on "upper back, right thigh, and right lower leg" gives 0.842. Considering practical usage, especially the ease of attaching sensors to patients, we conclude that the suitable number of sensors is three and the suitable combination of positions is "right thigh, back waist, and right lower leg" for recognizing rehabilitation motions.

The accuracy when DNN was used was inferior to the accuracy when using other classifiers. It is highly possible that this can be improved by parameter tuning such as the number and combination of nodes, layer type and activation function. However, there is no uniform method for parameter tuning. Accordingly, it is necessary to tune while looking at the feature of sensor data. Furthermore, CNN was used as a method of using DNN, there is a possibility that accuracy can be improved by using another method such as RNN.

Although recognizing ten rehabilitation motions was suggested by real rehabilitation practitioners, there are many more types of rehabilitation motions. As a future work, we are going to conduct similar experiments with an increased number of rehabilitation motions.

Acknowledgements We thank the National Center for Geriatrics and Gerontology, Japan for their advice and cooperation to the experiment.

References

ATR-Promotions. TSND121/151. ATR-Promotions [Online]. http://www.atr-p.com/products/TSND121.html. Accessed 24 Dec 2017

Gao L, Bourke AK, Nelson J (2014) Evaluation of accelerometer based multi-sensor versus single-sensor activity recognition systems. Med Eng Phys 36(6)

González-Villanueva L, Cagnoni S, Ascari L (2013) Design of a wearable sensing system for human motion monitoring in physical rehabilitation. Sensors 13(6)

Ikegaya T, Ooi S, Sano M (2016) Cooking action recognition based on first person vision for cognitive rehabilitation. In: Proceedings of the 78th national convention of IPSJ

Komukai K, Ohmura R (2018) Exploring the number and suitable positions of wearable sensors in automatic rehabilitation recording. In: 6th international workshop on human activity sensing corpus and application (HASCA)

Minamoto H, Sano M, Ooi S (2016) A human action recognition method for cooking rehabilitation. In: Proceedings of the 78th national convention of IPSJ, 2016

Mori I, Takahasi T, Hamasaki M, Shogo Y (2015) Development of basic movement scale (BMS) version 1: a new measure of basic movement capacity. Phys Ther Res 42(5) (in Japanese)

Pannurat N, Thiemjarus S, Nantajeewarawat E, Anantavrasilp I (2017) Analysis of optimal sensor positions for activity classification and application on a different data collection scenario. Sensors 17(4)

Vincent P, Larochelle H, Bengio Y, Manzagol P-A (2008) Extracting and composing robust features with denoising autoencoders. In: Proceedings of the ICML

Chapter 2
Identifying Sensors via Statistical Analysis of Body-Worn Inertial Sensor Data

Philipp M. Scholl and Kristof Van Laerhoven

Abstract Every benchmark dataset that contains inertial data (acceleration, rate-of-turn, magnetic flux) requires a thorough description of the datasets itself. This description tends often to be unstructured, and supplied to researchers as a conventional description, and in many cases crucial details are not available anymore. In this chapter, we argue that each sensor modality exhibits particular statistical properties that allow to reconstruct the modality solely from the sensor data itself. In order to investigate this, tri-axial inertial sensor data from five publicly available datasets are analysed. We found that in particular three statistical properties, the *mode*, the *kurtosis*, and the *number of modes* tend to be sufficient for classification of sensor modality—requiring as the only assumption that the sampling rate and sample format are known, and the fact that that acceleration and magnetometer data is present in the dataset. With those assumption in place, we found that 98% of all 1003 data points were successfully classified.

2.1 Introduction

For a benchmark to be properly usable, a significant amount of information about the data it holds is needed: the sampling (and frame) format, the recording rate, the number of axes, the position at the observed person's body and the sensor modality, to name just a few crucial ones. This collection of *meta-information* is usually supplied with the sensor data itself, either in a (semi-)structured external file, in

P. M. Scholl (✉)
University of Freiburg, Freiburg, Germany
e-mail: pscholl@informatik.uni-freiburg.de

K. Van Laerhoven
University of Siegen, Siegen, Germany
e-mail: kvl@eti.uni-siegen.de

N. Kawaguchi et al. (eds.), *Human Activity Sensing*,
Springer Series in Adaptive Environments,
https://doi.org/10.1007/978-3-030-13001-5_2

the header of the data set, or as a well-known convention. This meta-information's structure and level of detail are highly valuable: Misinterpretation of the data, as a result of missing or incorrect meta-information, can have a strong and negative impact on the subsequent performance. Furthermore, a slow and cumbersome recovery by a human expert is needed in case the meta-information is not presented in a machine-interpretable format. In this paper, we present extend and elaborate on a study presented in Scholl and van Laerhoven (2017) that investigates to which extent the *sensor modality* in a dataset can be recovered from invariant statistical properties of the sensor data itself. Motivations for such a method are diverse and numerous: (1) It enables snooping of wireless packets, without info about the sensor modality, (2) Revisiting a collected dataset is made possible, while info about sensor modality might have gotten lost or hidden, (3) It can aid in the generation of black box data or in the reconstruction of meta-data from sensor data, or it can be used (4) in cases where large amounts of data prohibit simply visualizing the data does not scale. We assume in the following that only the sample format, the number of axes, and the sample rate of the sensors are known beforehand, with the scale and other calibration factors unknown: This way, meta-information of activity recognition datasets could then be (automatically) verified, or even rendered redundant.

2.2 Related Work

There are several use cases for the automatic generation of (parts of) a dataset's meta-information. Providing structured meta-information enables datasets to be picked by search engines (Google Inc. 2017), for instance. In the absence or partial availability of this information, an automatic identification of sensor modality would provide a basic starting point that otherwise would require manual inspection by an expert. Webcrawlers could partially abandon *correctly* specified meta-information. Opportunistic sensing (Kurz et al. 2012), i.e. situations where the sensor type is not known beforehand, would be another application area: While proper data management practises could alleviate these situations, and are arguably more straightforward, an error in such manually defined data is often found much later in the process. Manually reconstructing meta-information in a sensor data stream is then hard to scale to large data collections, as inspection by a human expert is required. A second system, which identifies modality from data directly, could also provide additional safety. Such a type of quality control would allow to automatically check whether datasets were correctly documented or if there might be errors in the collection or documentation process, for example when uploading into a public dataset repository.

Human activity recognition applications are build with inherent assumptions about the data retrieved from the wearable inertial sensors. Properties such as the placement variations or body locations are assumed to be known, even though they directly influence the recognition performance. Kunze and Lukowicz (2014) have demonstrated such influences, and have provided several techniques to mitigate the resulting effects. They used location-independent features, adding location to the classifica-

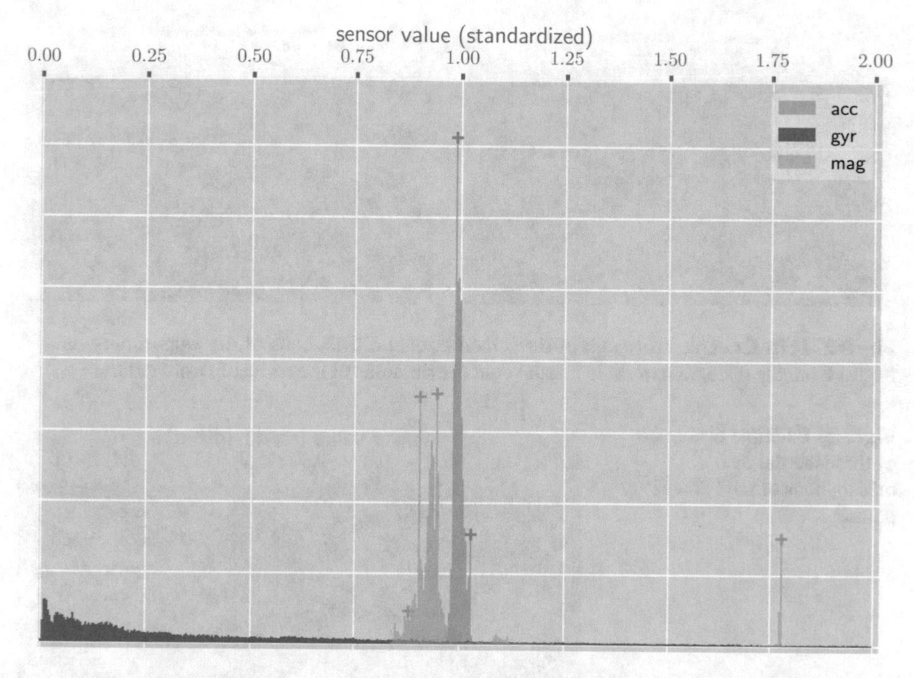

Fig. 2.1 Example histogram of three inertial data distributions of a CMU Kitchen dataset. Note that the concentration of the gyroscope data is around zero, as well as that the concentration of the acceleration data is around its mean. The larger number of modes for magnetometer data is visible, as well. Identified modes on the distribution are marked with a plus sign

tion task or estimating location from long-running recordings (Kunze and Lukowicz 2014).

Similarly to this latter option, we look in this chapter at the properties of different sensor modalities over longer time periods to estimate the sensor's modality. Whether sensor data arose from an accelerometer, gyroscope or a magnetometer is usually stored as non-standardized meta-information, but also strongly influences the recognition performance if incorrect. Hammerla and Kirkham (2013) introduced the empirical cumulative distribution function (ECDF) as a mean to capture the statistical properties of acceleration data, while also serving as a feature reduction method. Inspired by this approach, we here characterize *invariant* statistical features from the dataset that capture the properties of particular inertial sensor modalities.

A system to correlate known data streams to unknown ones and subsequently propagate their meta-data is described in Hartmann et al. (2013). In contrast to our proposed approach, it does require prior knowledge in the form of known sensor data, whereas we present a ruleset that can be readily applied. Our proposed method is, however, limited to inertial sensor data only (Figs. 2.1, 2.2 and 2.3).

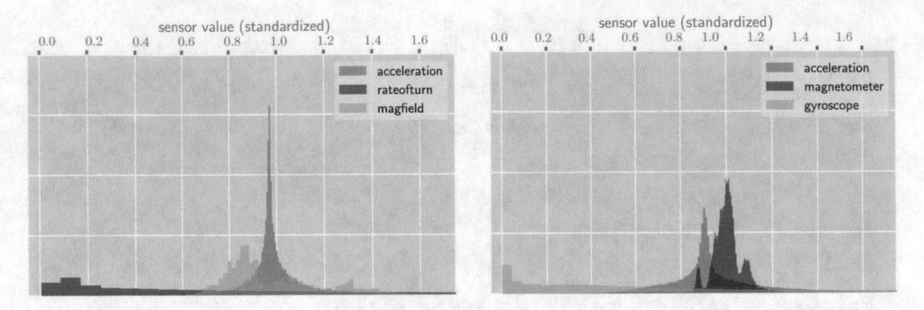

Fig. 2.2 Left: Example histogram of three inertial data distributions of the Opportunity dataset. Right: Example histogram of three inertial data distributions of the mHealthDroid dataset

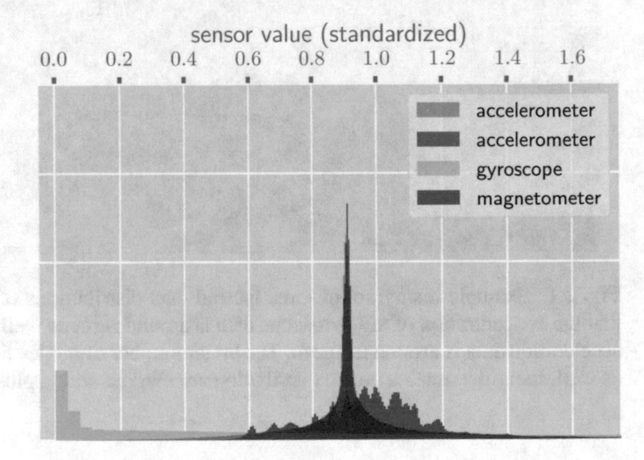

Fig. 2.3 Example histogram of three inertial data distributions of the Pamap2 dataset

2.3 Datasets

Publicly available datasets for activity recognition that include all three inertial sensor modalities, while optionally mounted at different body positions were selected and converted into a common data format using the method presented in Scholl and van Laerhoven (2016) to simplify their usage. The datasets used are:

CMU Kitchen (De la Torre et al. 2008) This dataset contains inertial data from sensors attached to multiple body locations, including the arms, the legs and the back of study participants. Even though two inertial capture systems were recorded, only the system with the wireless sensors, recording at 125 Hz, was used. To balance with the other datasets, a subset of 3 h and 26 participants was extracted for further analysis.

ICS Forth ADL (Karagiannaki et al. 2016) This benchmark dataset contains motion data from the thigh, ankle, torso and wrists. The measurements were taken at 50 Hz. 15 participants were recorded for a total of 4.5 h executing activities of daily living.

Pamap2 (Reiss et al. 2013) This dataset contains motion data of 8 participants executing activities of daily living at the hand, chest and ankles. In total 8 h were recorded. Inertial data was recorded at 100 Hz, using two acceleration sensors (at different scales), one magnetometer and one gyroscope.

Opportunity (Roggen et al. 2010) This set contains a whole-body inertial motion recording of daily living activities. A subset of 4 participants (with video recordings) contributed 15 data points each. In total 8 h of data recorded at 30 Hz was selected for the analysis.

mHealthDroid (Banos et al. 2014) has recorded 12 activities of daily living. Shimmer sensor nodes that were sampled at 100 Hz provided the inertial data used in this chapter. In total 6.5 h were used for analysis.

2.4 Pre-processing and Analysis of Sensor Data Properties

Wearable activity sensors differ in various aspects, and in order to estimate the sensor from the data, a few transformations are first needed. In order to simplify the overall analysis, only the *magnitude* of sensor readings is used instead of its vector form. This is achieved by applying the L2-norm to each sensor reading, which also renders all subsequent calculations *orientation-independent*. Since sensor streams might be scaled differently, e.g. the two acceleration streams in the pamap dataset were recorded with 6 g and 16 g range respectively, standardization is required. *Scale standardization* of the data is achieved by dividing the data points by the mean of the sensor data, which allows data to be compared. Additionally, data was also put through a lowpass filter using a cutoff frequency of 2 Hz. The last assumption that we take is that the sensor is usually at rest when worn on the human body. This way, we can now take a closer look at the properties of the transformed sensor streams, which in the following is referred to as d. The sensor streams we will focus on are the typical inertial sensors used in wearable activity recognition.

2.4.1 Detecting the Gyroscope Data

For gyroscopes, no rotation is measurable when the body it is attached to is at rest. The rate-of-turn for a body, as measured by a gyroscope, is therefore in most activity recognition datasets near zero most of the time. This knowledge can be employed to identify the gyroscope sensor modality by introducing the following simple rule:

$$mode(d) \approx 0 \Leftrightarrow gyr \tag{2.1}$$

This rule can be interpreted as: If the most common magnitude (also called the mode) of the sensor's data is near zero, then the sensor is likely a gyroscope. Vice versa, when we know the sensor is a gyroscope, we can expect the mode to be near zero. It is important to note that this rule only holds after the gyroscope data has been baseline-corrected.

2.4.2 Detecting the Accelerometer Data

The accelerometer's mode of magnitude corresponds to the strength of earth's gravitational field when no external motion is present for the sensor. This field strength is defined as $g = 9.81\ \mathrm{m\ s^{-1}}$, so we can formulate this as: $mode(d_{acc}) \approx a * g$, where $mode(d_{acc})$ is the most commonly measured value, and a an unknown scale factor applied to the data. If a would be known, the accelerometer data could be readily identified as such by comparing the data to earth's gravitation. As data was lowpass-filtered, however, the mean magnitude of acceleration corresponds to g as well, i.e. $\tilde{d}_{acc} \approx a * g$. Due to standardization, we can thus formulate a rule for acceleration:

$$acc \Rightarrow mode(d) \approx 1 \tag{2.2}$$

By applying this rule to the scatter plot depicted in Figs. 2.4 and 2.5, we can reveal why this is only a necessary condition; magnetometer data also fulfills this condition. A sufficient condition can be formulated for a subset of the overall accelerometer data, when including the kurtosis:

$$acc \Leftrightarrow mode(a) \approx 1 \wedge Kurt(d) > \alpha \tag{2.3}$$

Standardization is crucial for this condition, and relies on the assumption that the sensor is constantly accelerated by earth's gravitation. Other accelerations, due to limb movement for example, are only transient. Datasets which mostly contain strong movements, e.g. running or stirring as exemplary activities from the analysed data, will likely break this assumption. This is however tested with the mHealth, parts of the Opportunity and the Pamap dataset, which all contain sequences of strong, continuous motion.

2.4.3 Detecting the Magnetometer Data

Figure 2.4 shows that some magnetometer readings exhibit a mode and kurtosis that is indistinguishable from accelerometer data. However, fluctuations in the measured magnetic field are more distinct than fluctuations of the gravity field. The respective distribution therefore is not uni- but multi-modal. This means there are multiple peaks, while the accelerometer distribution is rather "smooth" (cf. Fig. 2.1). A mode larger than the mean (or 1 in the standardized dataset), and a smaller kurtosis can indicate this:

$$mag \Leftarrow Kurt(d) < \beta \wedge mode(d) \geq \gamma \tag{2.4}$$

Whether such strong fluctuations are contained in the dataset depends on the experiment's condition. By proper choice of β a subset of magnetometer data can be sufficiently identified. The smaller kurtosis can be explained due to the fact that

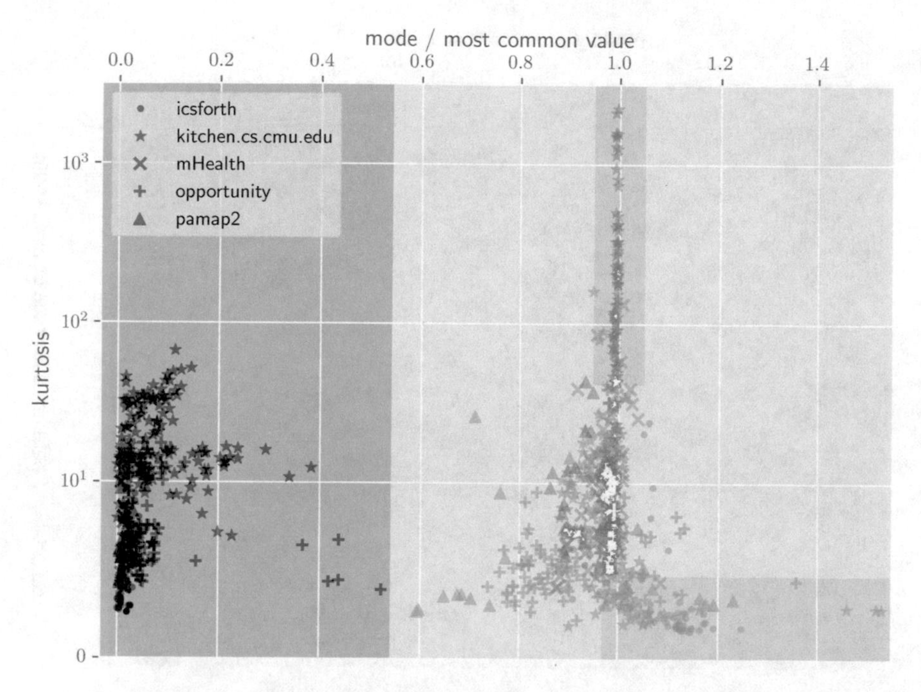

Fig. 2.4 Scatter plot of a possible feature set for the sensor modality detection. Shown is the mode of the histogram (using 128 equal-sized bins), or the most occurring value, versus the kurtosis of the data or the difference between the mean number of modes at the same limb and number of modes of one sensor stream. The decision threshold is shown as a highlighted layer

magnetometer is often further spread out, and does not exhibit a strong concentration point. In contrast, acceleration data has a strong concentration and its kurtosis is higher.

2.4.4 Distinguishing Acceleration Versus Magnetometer Data

One more aspect that needs to be solved is whether sensor streams that fulfill neither (2.3) nor (2.4) as conditions can still be identified. Or in other words, to find out what sensor the data stems from when a decision between acceleration or magnetic flux, based solely on kurtosis and mode, cannot be made. One observation that can be made about these cases, as well as the already identifiable cases, is that the number of modes for magnetometer is larger than the ones for acceleration data. Estimation of number of modes is achieved by adequately parameterized *peak detection* on the histogram. For a given stream, we designate the number of modes with p, as a shorthand for the number of peaks. However, streams have to be compared pairwise, i.e. magnetometer and accelerometer must have observed the same motion.

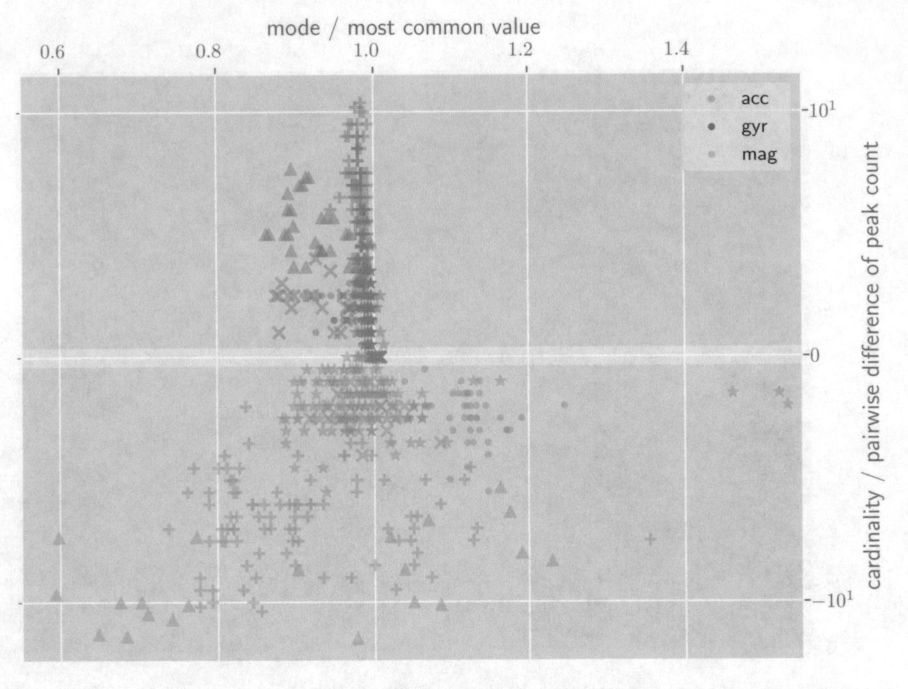

Fig. 2.5 Scatter plot of a possible feature set for the sensor modality detection, showing that not all cases can be identified with mode and kurtosis only. The decision threshold is shown as a highlighted layer

Therefore, let $\tilde{p} = \langle p \rangle$ designate the mean number of modes of correlated sensor streams, then we can formulate the following condition:

$$d \Leftrightarrow \begin{cases} acc, & \text{if } \tilde{p} - p > 0.5 \\ mag, & \text{if } \tilde{p} - p < -0.5 \\ unknown, & \text{otherwise} \end{cases} \tag{2.5}$$

Combined with (2.1) condition this allows to identify all sensor modalities, iff a correlated magnetometer and acceleration stream is to be distinguished.

2.5 Identification Ruleset

To finally identify the source of the data stream, this section presents the decision ruleset in this section. We define the *partial* ruleset as requiring solely the conditions (2.3) and (2.4), so we can define the *full* ruleset in conjunction with the pair-wise condition (2.5), as the following equation:

$$d \Leftrightarrow \begin{cases} gyr, & \text{if } m < 0.5 \\ acc, & \text{else if } k \geq 42 \text{ and } 0.95 < m \leq 1.05 \\ mag, & \text{else if } k < 4.3 \text{ and } 0.97 < m \\ acc, & \text{else if } \tilde{p} - p > 0.5 \\ mag, & \text{else if } \tilde{p} - p < -0.5 \\ unknown, & \text{otherwise} \end{cases} \tag{2.6}$$

where $m = mode(d)$ designates the mode of the data, $k = Kurt(d)$ its kurtosis, p the total number of modes and $\tilde{p} = \langle p \rangle$ the mean number of peaks of correlated data streams contained in one dataset.

The data is pre-processed in several passes, using particular constants and decision thresholds in these passes that were empirically determined in (2.6). The data needs first to be low-pass filtered to reject all frequencies above 2 Hz. Furthermore, to reduce scaling effects, a standardization pass is applied by dividing by the mean of each stream. The mode is then determined from a histogram of 512 equal-sized bins, ranging from 0 to 2. The peak detection parameters were set to a minimum peak height of $0.01 * m$, minimum distance of $5bins$ and a minimum neighbor difference of $0.008 * m$.

2.6 A Discussion of Results and Limitations of Our Approach

In the above rulesets and analysis, it should be noted that a few limitations make our results not entirely generic. We analyzed a set of 1003 data sequences that lasted anywhere between 7 min and 1 h were analyzed. All previously-discussed inertial sensor modalities are included, mostly positioned at the lower arm (61%), the upper body (20%) and the legs (19%). Data is scaled differently for each included dataset, showing that the proposed ruleset is independent of particular scale. Similarly, the sampling rates for each dataset differ. The *full* ruleset allows to identify 98% of all cases, while 2% remain for manual inspection, as can also be seen in the confusion matrices in Table 2.1. If streams can not be compared pair-wise, the *partial* ruleset can still identify 51% of all cases, of those less than 1% are wrongly classified, while the remaining require manual inspection.

As the above threshold and features were designed from and tested on the same set of data points, one could argue that the proposed ruleset will not generalize to other, unseen sensor time series and datasets. Or in other words, we might just observe an over-fitted solution to this classification task. This could be investigated by maximizing the set of parameters (lowpass cutoff frequency, peak detection parameters, thresholds of (2.6) ...) on leave-one-dataset-out splits for detection accuracy. In a worst case scenario, then, there is no set of parameters whose performance will be doing well for all datasets, i.e. there is no generalizing set of parameters across all

Table 2.1 The confusion matrices for the sensor identification with the *full* (table on the left) and *partial* (on the right) rulesets. The full ruleset fails to identify the correct sensor for 2% of the analysed sensor streams, but correctly identifies them for further manual inspection

	–	acc	gyr	mag		–	acc	gyr	mag
–		16		6	–		276		205
acc		339			acc		78		6
gyr			324		gyr			324	
mag				318	mag		1		113

splits. In a best case scenario, a *single* set of parameters can be found that performs equally well across all splits.

Figure 2.4 allows some insights how threshold parameters could be chosen differently, if one dataset would have been left out. However, not all parameters are chosen based on these data points alone (what a machine learning approach would roughly do): (1) the mode threshold is based on the insight that gyroscope data is concentrated near zero, (2) the pair-wise peak threshold follows the observation that the magnetometer distribution exhibits more modes. This is the case for 98% of the observed data points. The latter observation has examples in multiple datasets, as is visible in Fig. 2.4, partially ruling out an over-fit. A cross-validated *automatic* choice of parameters would reveal if the opposite was true, in a formal way. Here we merely report a single set of parameters that worked—the probability of an over-fitting is low, but a better choice of parameters that maximizes the decision boundaries may be possible.

One of the research questions that hasn't been solved in this chapter is how much data is needed to achieve a reliable estimation on the sensor that produced the data. For each dataset, the full data was used each time for feature computation, but varying the amount of data would provide more detailed insights into the amount and nature of the required data. This, however, was not attempted to avoid over-estimating the quality of the decision. A further hurdle is that standardization by dividing by the mean can be problematic if the sensor was asymmetrically driven into saturation, for instance, when the magnetometer was exposed to unipolar magnetic interference. The mode could in such cases be nearer to zero, yielding an incorrect classification. A possible solution to this problem could be to remove outliers by applying filters.

2.7 Conclusions

This chapter has proposed and described a method to statistically describe one of the most common type of human activity recognition datasets, those containing body-worn inertial data from sensors consisting of accelerometers, gyroscopes, or magnetometers. With this method, by applying the conditions formulated in Eq. (2.6) on an inertial sensor data stream, it is possible to classify its modality, if the sensor was worn on the human body. We have shown through analysis of the data properties that the sensor modality can be estimated from its sensor data, allowing an automatic quality control of available meta-data and the data itself, as well as opportunistic sensing, and interpretation of datasets with unstructured or faulty meta-data. The mode of the distribution for data from gyroscope sensors has shown to be a strong indicator (see Fig. 2.4), so deciding whether a sensor was a gyroscope will likely generalize to multiple settings. The distinguishing between acceleration and magnetometer data is however more challenging. As shown in our experiments, only the pair-wise comparison of peak count has provided a clear indication. Part of the problem is likely due to the presence of environmental fluctuations present in the magnetometer data, which tends to be smaller when the data contain less activity and motion. The presented approach was shown to be able to classify 98% out of 1003 instances in 5 different human activity datasets.

Acknowledgements and Reproducibility We would like to thank the authors of the datasets that were used in this chapter, for making them available for public use. All code and intermediate results for the proposed system are available online at the following website: https://github.com/pscholl/imustat.

References

Banos O, Garcia R, Holgado-Terriza JA, Damas M, Pomares H, Rojas I, Saez A, Villalonga C (2014) mHealthDroid: a novel framework for agile development of mobile health applications. In: International workshop on ambient assisted living. Springer, pp 91–98

De la Torre F, Hodgins J, Bargteil A, Martin X, Macey J, Collado A, Beltran P (2008) Guide to the Carnegie Mellon University multimodal activity (CMU-MMAC) database. Robotics Institute

Google Inc. (2017) Google developer guides to dataset curaton (2017). https://developers.google.com/search/docs/data-types/datasets

Hammerla N, Kirkham R (2013) On preserving statistical characteristics of accelerometry data using their empirical cumulative distribution. In: Proceedings of the 2013 international symposium on wearable computers, pp 65–68

Hartmann M, Bauer A, Blanke U (2013) Method and system for sensor classification. US Patent 7062320

Karagiannaki K, Panousopoulou A, Tsakalides P (2016) A benchmark study on feature selection for human activity recognition. In: Proceedings of the 2016 ACM international joint conference on pervasive and ubiquitous computing: adjunct. ACM, pp 105–108

Kunze K, Lukowicz P (2014) Sensor placement variations in wearable activity recognition. In: Proceedings of the 2014 international symposium on wearable computers, vol 13, pp 32–41

Kurz M, Holzl G, Ferscha A, Calatroni A, Roggen D, Troster G, Sagha H, Chavarriaga R, Millan JR, Bannach D, Kunze K, Lukowicz P (2012) The opportunity framework and data processing ecosystem for opportunistic activity and context recognition. Int J Sens Wirel Commun Control 1(2):102–125

Reiss A, Hendeby G, Stricker D (2013) Confidence-based multiclass adaboost for physical activity monitoring. In: Proceedings of the 2013 international symposium on wearable computers. ACM, pp 13–20

Roggen D, Calatroni A, Rossi M, Holleczek T, Kilian F, Tr G, Lukowicz P, Bannach D, Pirkl G, Ferscha A, Doppler J, Holzmann C, Kurz M, Holl G, Creatura M, Mill R (2010) Collecting complex activity datasets in highly rich networked sensor environments. In: International conference on networked sensor systems

Scholl PM, Van Laerhoven K (2016) A multi-media exchange format for time-series dataset curation. In: Proceedings of the 2016 ACM international joint conference on pervasive and ubiquitous computing: adjunct. ACM, pp 715–721

Scholl PM, van Laerhoven K (2017) On the statistical properties of body-worn inertial motion sensor data for identifying sensor modality. In: Proceedings of the 2017 ACM international symposium on wearable computers—ISWC 17. ACM Press. https://doi.org/10.1145/3123021.3123048

Chapter 3
Compensation Scheme for PDR Using Component-Wise Error Models

Junto Nozaki, Kei Hiroi, Katsuhiko Kaji and Nobuo Kawaguchi

Abstract There is an inherent problem of error accumulation in Pedestrian Dead Reckoning (PDR). In this chapter, we introduce a PDR error compensation scheme based on the assumption that can obtain sparse locations. Sparse locations are discontinuous locations obtained by using an absolute localization method or passage detection devices (ex. RFID tag, BLE beacon, Spinning Magnet Marker). Our proposal scheme focuses on being able to install anywhere in the indoor environment. In our scheme, we define error models that represent errors in PDR, including moving distance error and orientation change error. We apply the error models to counteract the error that occurs in PDR estimation. Moreover, the error models are tuned each time when a sparse location is measured. As a result, the proposed scheme improves the position error rate by approximately 10% and the route distance error rate by approximately 7%. In addition, we discuss the effectiveness of our scheme by each test route for future consideration.

J. Nozaki (✉)
Graduate School of Engineering, Nagoya University, Furo-cho, Chikusa-ku, Nagoya, Aichi 464-8603, Japan
e-mail: nozaki@ucl.nuee.nagoya-u.ac.jp

K. Hiroi
Faculty of Graduate School of Engineering, Nagoya University, Furo-cho, Chikusa-ku, Nagoya, Aichi 464-8603, Japan
e-mail: k.hiroi@ucl.nuee.nagoya-u.ac.jp

K. Kaji
Faculty of Information Science, Aichi Institute of Technology, 1247 Yachigusa, Yakusa-cho, Toyota, Aichi 470-0392, Japan
e-mail: kaji@aitech.ac.jp

N. Kawaguchi
Faculty of Graduate School of Engineering, Institutes of Innovation for Future Society, Nagoya University, Furo-cho, Chikusa-ku, Nagoya, Aichi 464-8603, Japan
e-mail: kawaguti@nagoya-u.jp

N. Kawaguchi et al. (eds.), *Human Activity Sensing*,
Springer Series in Adaptive Environments,
https://doi.org/10.1007/978-3-030-13001-5_3

3.1 Introduction

Various indoor localization methods have been proposed. These are roughly divided into absolute localization and relative localization. In absolute localization, the absolute coordinates of the localization target are estimated by using the device arranged in the environment. As a typical method, there are methods using Wi-Fi access points (Farshad et al. 2013; Ferris et al. 2007), RFID tags (Oberli et al. 2010) or BLE (Bluetooth low energy) beacons (Faragher and Harle 2015; Ciabattoni et al. 2017). In relative localization, the location is estimated by accumulating the movement amount from the past location by using inertial sensors attached to the localization target. In particular, the method for pedestrians is called PDR, and various methods using smartphones have been proposed. Relative localization is superior to absolute localization requiring to install additional devices in terms of being easily available.

Although PDR is easily available, there are problems in accuracy. Estimation errors occur in step detection, step length estimation and orientation estimation which are components of PDR, and localization error accumulates over time. To compensate for these errors, it is necessary to use different information. Map matching (Shin et al. 2012) using spatial constraints can be given as an example.

In our previous research (Nozaki et al. 2017), we proposed a compensation scheme using sparse location. An idea of our method is arranging devices which are used for absolute localization sparsely, and use them as reference points of compensation. It is necessary to densely when used for absolute localization method, however, our scheme requires minimal installing cost. RFID tags, BLE beacons or spinning magnet markers (Takeshima et al. 2015) can be given as examples of the device obtaining sparse location. This scheme can be deployed in a broad area where map matching is difficult to use.

To compensate for errors, we define the error models for each cause of the error. They are moving distance error model, orientation change error model and drift error model. Updating the error models using each sparse location obtained, we attempt to improve accuracy. In this chapter, we describe our compensation scheme and conduct additional consideration for individual test route. That lead to future consideration of compensation schemes.

3.2 Related Work

There are several approaches to solving the problem of accumulation error in PDR. One approach is to reduce the estimation error by combining many sensors. Most PDR method uses accelerometer and gyroscope. In addition to this, a method using magnetometer (Kang and Han 2015) or barometer (Krach and Roberston 2008) has been proposed. Barometers are also used to extend estimation 3D localization in multiple floor situations (Zheng et al. 2016; Li et al. 2014). These methods are still used only relative localization, therefore it does not become a fundamental solution. For this reason, some schemes that are combined with other localization methods have been proposed. For example, a scheme that combines PDR with Wi-Fi fingerprinting (Chang et al. 2015) or magnetic field fingerprinting (Ban et al. 2015). These schemes estimate location by combining information of Wi-Fi signal strengths or magnetic sensor values with moving distances and orientation changes estimated by PDR. However, these schemes require collecting environmental information and creating a map fingerprint in advance. When we collect environmental information, the whole area should be collected. Therefore, the measurement, maintenance and updating costs are high.

Another approach is to compensate the estimation. For example, a scheme that adjusts the user's step length utilizing GPS (Miyake and Arai 2013) or compensates for estimation using map-matching utilizing building structural information. The former scheme adjusts the step length from step counts obtained by step detection utilizing an accelerometer and the moving distance obtained from a GPS. This scheme can reduce the moving distance error. However, because it requires calibration by walking outdoors before indoor localization, it is difficult to use. The latter scheme compensates for the user's step length and orientation change utilizing turn detection and the corner information of a map. However, requiring turn information at corners is difficult to use in a broad area such as an airport lobby.

In the conventional compensation scheme for PDR, the situations that can apply for estimation are limited. For practical compensation scheme for PDR, it is required that the method that is easily installed on anywhere and is costless.

3.3 Compensation Scheme Proposal

In this section, we describe the scheme of our previous research. Our scheme utilizes high precision locations in some specific points, we call such points sparse locations. We assume that sparse locations are obtained by devices which are used for absolute localization, for example, RFID tags, BLE beacons or spinning magnet marker. These devices can be deployed in optical locations also in a broad area, and this allows us to extend the applicable situations of compensation.

In our scheme, we define the error models for moving distance, orientation change and drift angle. Parameters describing these error models are updated every time a sparse location is measured, and then they are utilized in PDR estimation. Figure 3.1 shows the whole system of the proposed scheme. As shown in Fig. 3.1, our scheme compensates for moving distance error after compensating for orientation change error, because the compensation for moving distance error uses the distance between a location estimated by PDR and a sparse location. This distance is changed by orientation change error compensation.

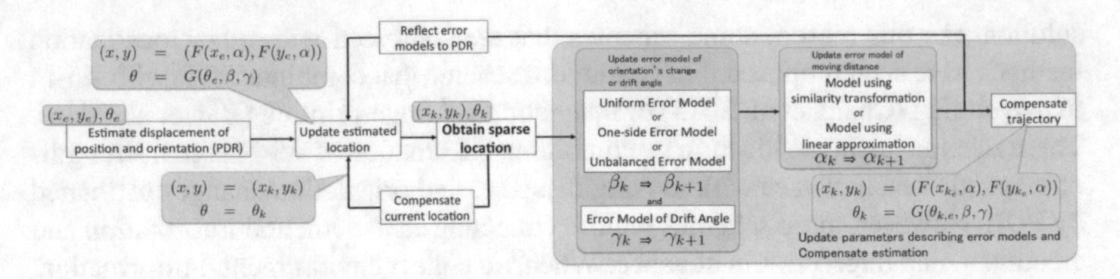

Fig. 3.1 System flow of the proposed scheme

Table 3.1 Symbols used in the section and their explanation "Moving Distance Error Model"

Symbol	Explanation
t_k	The time obtained at the kth sparse location
(x_k, y_k)	The coordinate of the kth sparse location
$(x_{k,e}, y_{k,e})$	The coordinate of the estimated location when the kth sparse location is obtained
(x_k^e, y_k^e)	The coordinate of the estimated location when compensation is not used
d_k	The error distance when compensation is not used
δt_k	$t_k - t_{k-1}$, the time from the $k - 1$th sparse location to kth sparse location
(x_e, y_e)	A coordinate of the estimated location
(x, y)	A coordinate of the compensated location
α	The parameter describing the moving distance error model

3.3.1 *Moving Distance Error Model*

In this research, we propose two moving distance error models. One uses a relationship between moving distance and moving distance error, and the other uses a relationship between elapsed time and moving distance error. A parameter describing the former is moving distance error generated per meter, and another parameter describing the latter is moving distance error generated per second. The descriptions of the symbols used in this section are summarized in Table 3.1.

3.3.1.1 Model Using Similarity Transformation

This model is based on the idea that moving distance error accumulates proportionally to the moving distance from the initial location. The distance from a previous sparse location to a new sparse location is exactly the same as the user's moving distance. Therefore, compensate $(x_{k,e}, y_{k,e})$ to be equal to (x_k, y_k) using a similarity transformation. First, we calculate the scale factor s. Then, we update the parameter describing this model by the following equations.

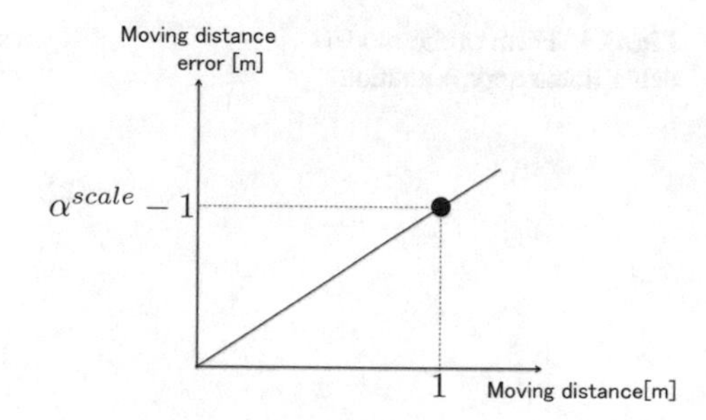

Fig. 3.2 Form of the model using similarity transformation

$$x=\sqrt{\frac{(x_k-x_{k-1})^2+(y_k-y_{k-1})^2}{(x_{k,e}-x_{k-1})^2+(y_{k,e}-y_{k-1})^2}} \tag{3.1}$$

$$\alpha_k^{scale}=s\alpha_k^{scale} \tag{3.2}$$

The initial value of α^{scale} is 1. $\alpha^{scale}-1$ represents the slope which sets the moving distance on a horizontal axis and the moving distance error on a vertical axis. As shown in Fig. 3.2, this parameter describes this similarity transformation model. Using this model, we apply compensation to PDR estimation. The estimated location (x_e, y_e) after the time (x, y) from the previous sparse location is compensated to (x, y) by the following equation.

$$\begin{pmatrix} x \\ y \end{pmatrix}=\begin{pmatrix} x_k \\ y_k \end{pmatrix}+\alpha^{scale}\begin{pmatrix} x_e-x_k \\ y_e-y_k \end{pmatrix} \tag{3.3}$$

3.3.1.2 Model Using Linear Approximation

This model is based on the idea that moving distance error accumulates proportionally to elapsed time. The parameter describing this model is updated by applying the linear approximation to sets of an elapsed time and a moving distance error. First, we calculate the moving distance error d_k using (x_k^e, y_k^e): the coordinates estimated by PDR that do not compensate for estimation by utilizing the error model. At this time, hold d_k and moving time δtk. Then, update the parameter describing this model by the following equations.

$$d_k=\sqrt{\left(x_k-x_k^e\right)^2+\left(y_k-y_k^e\right)^2} \tag{3.4}$$

$$\alpha_{slope}=\frac{\sum_{i=0}^{k}\Delta t_i d_i}{\sum_{i=0}^{k}t_i^2} \tag{3.5}$$

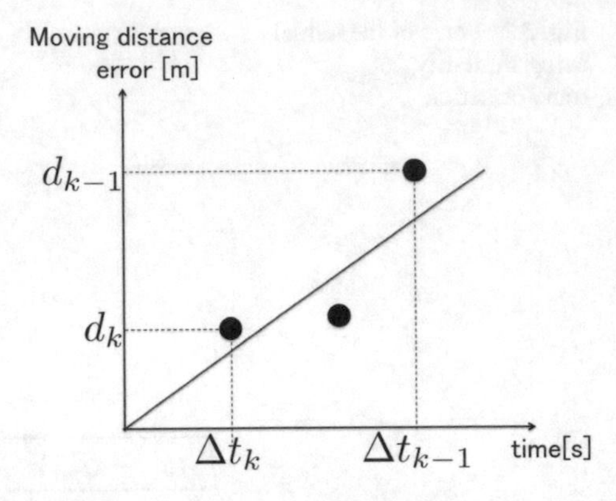

Fig. 3.3 Form of the model using linear approximation

α_{slope} represents the slope which sets the moving time on a horizontal axis and the moving distance error on a vertical axis. As shown in Fig. 3.3, this parameter describes this linear approximation model. Using this model, we apply compensation to PDR estimation. The estimated location (x_e, y_e) after the time τ from the previous sparse location is compensated to (x, y) by the following equations.

$$d_k = \sqrt{(x_e - x_k)^2 + (y_e - y_k)^2} \tag{3.6}$$

$$\begin{pmatrix} x \\ y \end{pmatrix} = \begin{pmatrix} x_k \\ y_k \end{pmatrix} + \frac{d_e + \tau\alpha_{slope}}{d_e} \begin{pmatrix} x_e - x_k \\ y_e - y_k \end{pmatrix} \tag{3.7}$$

3.3.2 *Error Model of Orientation Changing*

In this research, we propose three patterns for the orientation change error model that use a relationship between orientation change and orientation change error. The parameter, which describes these models, is the orientation change error generated per degree. We designed three patterns of the model because we consider that it is possible that orientation change error may occur at different rates when people turn left or turn right. The description of symbols used in this section is summarized in Table 3.2.

3.3.2.1 How to Calculate Orientation Change

For example, in the case in which the orientation of the kth sparse location is equal $k-1$, the value of θ_k cannot be determined uniquely but can take $2n\pi\,(n = 0, 1, 2, \ldots)$. Therefore, we calculate θ_k as follows. First, we calculate the orientation

Table 3.2 Symbols used in this section and their explanations "Orientation Change Error Model" and "Error Model of Drift Angle"

Symbol	Explanation
t_k	The time obtained at the kth sparse location
θ_k	The orientation change from the $k-1$th sparse location to kth one
$\theta^+_{e,k}, \theta^-_{e,k}$	The accumulation of estimated positive or negative orientation changing
$diff$	The number of updating parameters
θ_e	An estimated orientation change
θ	A compensated orientation change
β	The parameter describing the error model of orientation
γ	The parameter describing the error model of drift angle

change error by the following equation.

$$\theta_{error} = \left(\theta^+_{e,k} - \theta^-_{e,k}\right) - \theta_k \tag{3.8}$$

The initial value of θ_k is 0. Second, assume that the error of orientation change is in the range $\pi < \theta_{error} < \pi$, calculate θ_k by the following equation.

$$\begin{cases} \theta_k = \theta_k + 2\pi \ (\theta_{error} > \pi) \\ \theta_k = \theta_k - 2\pi \ (\theta_{error} < \pi) \end{cases} \tag{3.9}$$

Finally, execute the operation of equation (3.8) and (3.9) recursively until θ_{error} is in the range $\pi < \theta_{error} < \pi$.

3.3.2.2 Uniform Error Model

This model is based on the idea that each orientation change error may occur uniformly. First, we calculate $diff_{uniform}$: the ratio of the orientation change error to the orientation change. Then, update $\beta_{uniform}$, which is the parameter describing this model, by the following equations.

$$diff_{uniform,k} = \frac{\theta_{error}}{\theta^+_{e,k} + \theta^-_{e,k}} \tag{3.10}$$

$$\beta_{uniform,k} = \beta_{uniform,k-1} - diff_{uniform,k} \tag{3.11}$$

The initial value of $\beta_{uniform}$ is 0. $\beta_{uniform}$ represents the slope which sets the orientation change on a horizontal axis and orientation change error on a vertical axis. As shown in Fig. 3.4, this parameter describes this uniform error model. Using this model, we

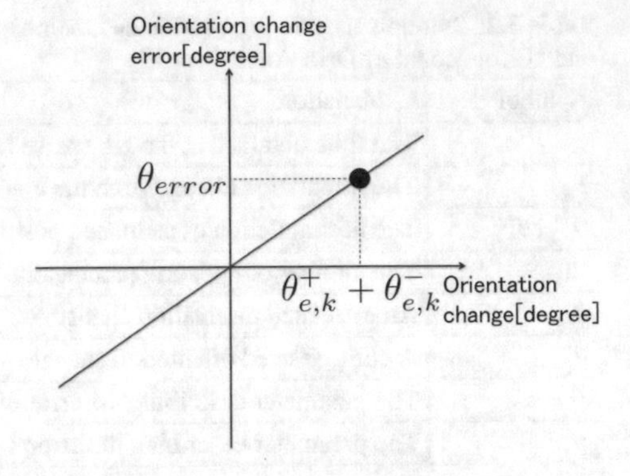

Fig. 3.4 Form of the uniform error model

apply compensation to the PDR estimation. The estimated orientation change θ_e from the previous sparse location is compensated to θ by the following equation.

$$\theta = \theta_e \left(1 + \beta_{uniform}\right) \tag{3.12}$$

3.3.2.3 One-Sided Error Model

This model based on the idea that all the orientation change errors may occur on one-side, either positive or negative. First, we calculate the ratio of orientation change error in the positive or negative direction orientation change $diff^{+}_{oneside}$ and $diff^{-}_{oneside}$. Then, we update $diff^{+}_{oneside}$ and $diff^{-}_{oneside}$, which are the parameters describing this model by the following equations.

$$\begin{cases} diff^{+}_{oneside,k} = \frac{\theta_{error}}{\theta^{+}_{e,k} - \theta^{-}_{e,k}} \\ diff^{-}_{oneside,k} = -diff^{+}_{oneside,k} \end{cases} \tag{3.13}$$

$$\begin{cases} \beta^{+}_{oneside,k} = \beta^{+}_{oneside,k-1} + diff^{+}_{oneside,k} \\ \beta^{-}_{oneside,k} = \beta^{-}_{oneside,k-1} + diff^{-}_{oneside,k} \end{cases} \tag{3.14}$$

The initial value of $\beta^{+}_{oneside}$ and $\beta^{-}_{oneside}$ are both 0. $\beta^{+}_{oneside}$ and $\beta^{-}_{oneside}$ represent slopes that set orientation change on a horizontal axis and orientation change error on a vertical axis. As shown in Fig. 3.5, this parameter describes this one-sided error model. Using this model, we apply compensation to PDR estimation. The estimated orientation change θ_e from the previous sparse location is compensated to θ by the following equation.

$$\begin{cases} \theta = \theta_e \left(1 + \beta^{+}_{oneside}\right) \ (\theta > 0) \\ \theta = \theta_e \left(1 + \beta^{-}_{oneside}\right) \ (\theta < 0) \end{cases} \tag{3.15}$$

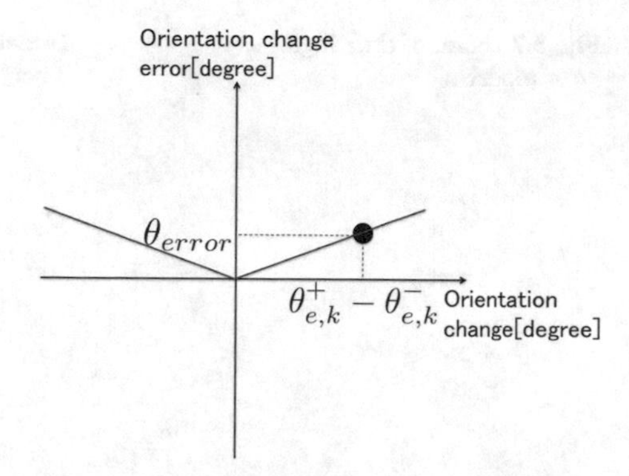

Fig. 3.5 Form of the one-sided error model

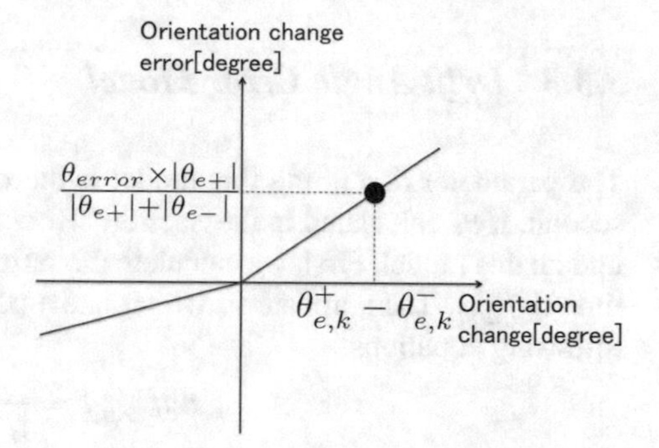

Fig. 3.6 Form of unbalanced error model

3.3.2.4 Unbalanced Error Model

This model is based on the idea that orientation change errors may occur at different rates in positive or negative direction orientation changes. First, we calculate the ratio of orientation change error to a positive or negative direction orientation change $diff^+_{unbalanced}$ and $diff^-_{unbalanced}$ by the Eq. 3.16. Then, we update $diff^+_{unbalanced}$ and $diff^-_{unbalanced}$, which are the parameters that describe this model in the same way as Eq. 3.14.

$$\begin{cases} diff^+_{unbalanced,k} = \frac{\theta_{error}}{\theta^+_{e,k}+\theta^-_{e,k}} \frac{|\theta^+_{e,k}|}{|\theta^+_{e,k}|+|\theta^-_{e,k}|} \\ diff^-_{unbalanced,k} = \frac{\theta_{error}}{\theta^+_{e,k}+\theta^-_{e,k}} \frac{|\theta^-_{e,k}|}{|\theta^+_{e,k}|+|\theta^-_{e,k}|} \end{cases} \tag{3.16}$$

The initial values of $diff^+_{unbalanced}$ and $diff^-_{unbalanced}$ are both 0. $diff^+_{unbalanced}$ and $diff^-_{unbalanced}$ represent slopes, which set the orientation change on a horizontal axis and the orientation change error on a vertical axis. As shown in Fig. 3.6, this parameter describes this unbalanced error model. Using this model, we apply compensation to PDR estimation. The estimated orientation change θ_e after the time τ from the previous sparse location is compensated to θ in the same way as in Eq. 3.15.

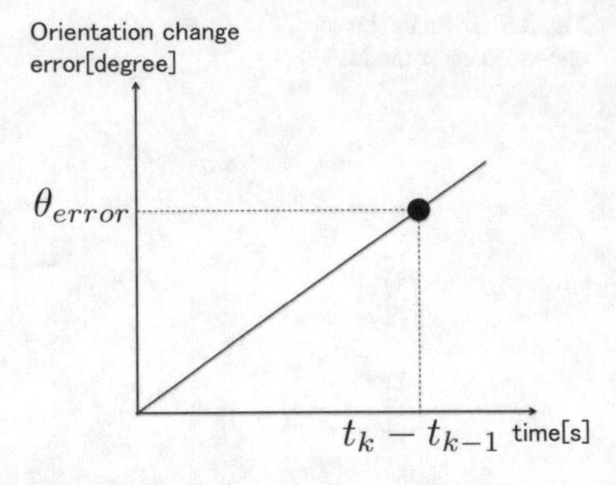

Fig. 3.7 Form of drift angle error model

3.3.3 *Drift Angle Error Model*

The parameter describing this model is the orientation change error generated per second. If θ_k calculated in the Section "How to Calculate Orientation Change" is 0, update this model. First, we calculate the ratio of orientation change error to elapsed time $\mathit{diff}_{\mathit{drift}}$. Then, update γ, which is the parameter describing this model, by the following equations.

$$\mathit{diff}_{\mathit{drift},k} = \frac{\theta_k}{t_k - t_{k-1}} \tag{3.17}$$

$$\gamma_k = \gamma_{k-1} + \mathit{diff}_{\mathit{drift},k} \tag{3.18}$$

The initial value of γ is 0. γ represents the slope which sets the elapsed time on a horizontal axis and the orientation change error on a vertical axis. As shown in Fig. 3.7, this parameter describes this drift angle error model. Using this model, we apply compensation to the PDR estimation. The estimated orientation change θ after the time τ from the previous sparse location is compensated to θ by the following equation.

$$\theta = \theta_e + \gamma\tau \tag{3.19}$$

3.4 Evaluation

For evaluating our proposed scheme, we collected pedestrian walking data from the Geospatial EXPO (2016) held at the National Museum of Emerging Science and Innovation. We created a situation that can obtain sparse locations around the four corners. As the sparse location, we utilized the locations obtained from a UWB (ultrawideband) localization system, which needs dedicated devices. However, we obtained sparse locations, which contained error of approximately 15–30 cm. More-

Table 3.3 Route information

Number of subjects	2
Number of routes	4
Place of smartphone	Back waist
Size of the area	23 m × 70 m

Table 3.4 Experiment information

	Length	Corner
A	120	3
B	160	7
C	215	19
D	120	3

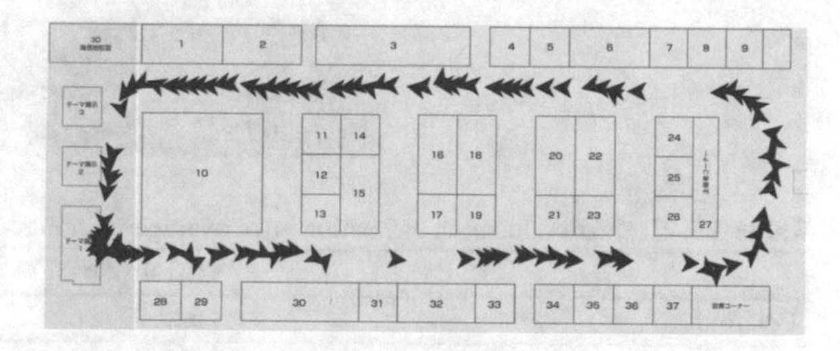

Fig. 3.8 Route A

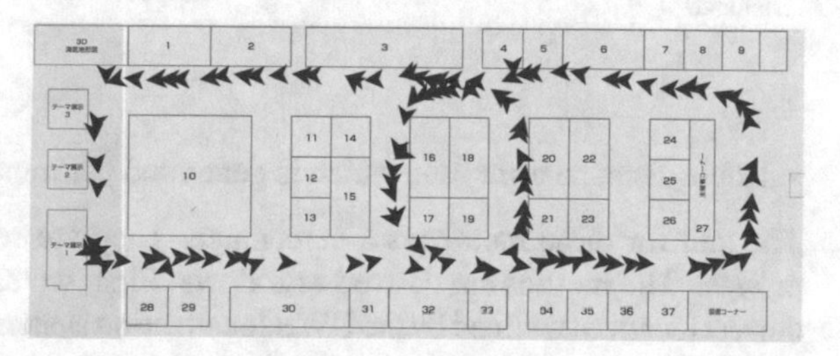

Fig. 3.9 Route B

over, we also utilized the locations obtained from the UWB localization system as the correct location to evaluate the proposed scheme.

Tables 3.3 and 3.4 and Figs. 3.8, 3.9, 3.10 and 3.11 show the information about the collected data and used routes. In the proposed scheme, the parameters describing the error models were updated every time a sparse location was measured as mentioned in the section "Compensation Scheme Proposal". Moreover, the previously estimated trajectory was compensated at the same time. This trajectory compensation is useful for the purpose of people flow analysis. However, trajectory compensation is meaningless for the purpose of navigation; real-time estimation is important for this purpose. Therefore, we evaluate post-compensation, which executes the previous trajectory compensation, and real-time compensation, which does not execute this.

We use following to evaluate the metric based on the reference, Abe et al. (2016).

- Position error rate, which is generated per second

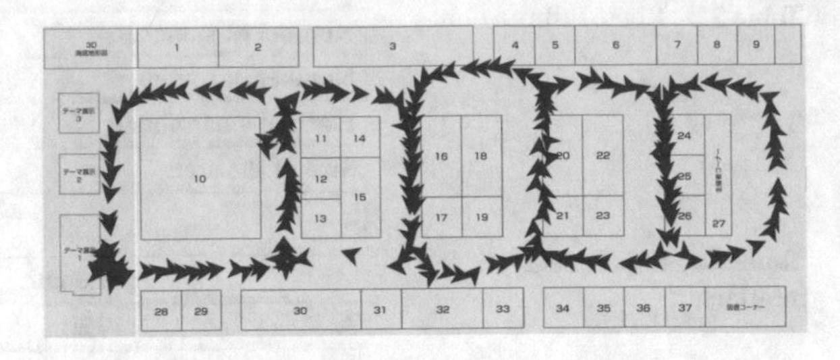

Fig. 3.10 Route C

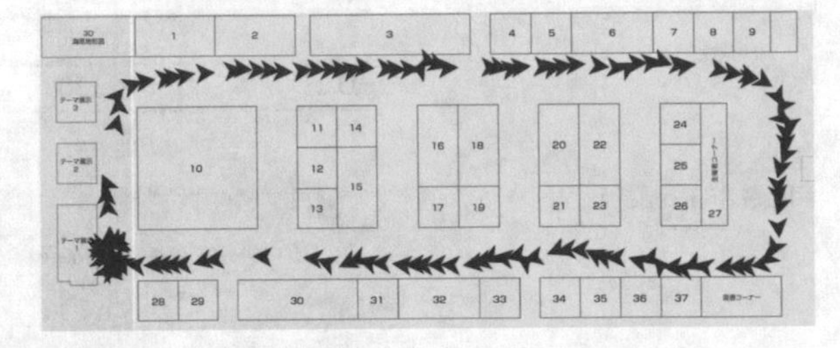

Fig. 3.11 Route D

Table 3.5 Evaluation location estimation error average without compensation

	Error rate (%)
Position	52.55
Distance	36.11

- Route distance error rate, which is generated per meter

The unit for these metrics is a percentage. They are represented by stand for m/s or m/m. The metrics are derived as follows. First, we calculate each error for every correct location obtained by the UWB localization system. Second, we create a scatter plot that sets the time or distance on the horizontal axis and error on the vertical axis by plotting all the calculated errors. Third, we calculate the slope of the regression line by using the least square estimation method. This slope is the evaluation metric. Finally, we calculate the average of the evaluation location estimation error for the four routes. Table 3.5 shows the average of the evaluation location estimation error for each second or meter without compensation.

3.4.1 Post-compensation

Tables 3.6 and 3.7 shows the average of the evaluation location estimation error. Post-compensation, the position error rate improved to approximately 10% when utilizing the similarity transformation model. Moreover, the route distance error rate improved to approximately 7% when utilizing the linear approximation model.

Table 3.6 Evaluation location estimation error average of post-compensation when utilizing the model using similarity transformation

	Without model (%)	Uniform (%)	One-side (%)	Unbalanced (%)
Position	17.19	11.78	10.39	10.85
Distance	16.50	7.70	7.96	7.01

Table 3.7 Evaluation location estimation error average of post-compensation when utilizing the model using linear approximation

	Without model (%)	Uniform (%)	One-side (%)	Unbalanced (%)
Position	17.01	15.53	14.63	13.49
Distance	6.40	7.21	7.24	7.01

Table 3.8 Evaluation location estimation error average of real-time compensation when utilizing the model using similarity transformation

	Without model (%)	Uniform (%)	One-side (%)	Unbalanced (%)
Position	27.43	28.65	31.35	29.47
Distance	13.94	15.33	14.36	15.09

Table 3.9 Evaluation location estimation error average of real-time compensation when utilizing the model using linear approximation

	Without model (%)	Uniform (%)	One-side (%)	Unbalanced (%)
Position	25.19	24.94	28.14	26.55
Distance	11.42	11.70	11.42	11.61

3.4.2 Real-Time Compensation

Tables 3.8 and 3.9 shows the average of the evaluation location estimation error. In real-time compensation, utilizing the linear approximation model tends to be acceptable rather than utilizing the similarity transformation model. Both evaluation metric values decrease by approximately 3%.

3.4.3 Discussion About the Moving Distance Error Model

Post-compensation, from the viewpoint of position error rate, utilizing the similarity transformation model tends to be acceptable rather than utilizing the linear approximation model. The best error rate when utilizing the similarity transformation model is 10.37%, and the best error rate when utilizing the other model is 13.49%. From the viewpoint of route distance error rate, it is the opposite. The best error rate when utilizing the similarity transformation model is 7.70%, and when utilizing the other model, it is 6.40%. This is because the similarity transformation model compensates the estimation when obtaining a sparse location so that it coincides with the sparse location. Therefore, the position error rate improves. Conversely, the linear approximation model, which uses the error distance per second, is reflected in the estimation of moving velocity. Therefore, the route distance error rate decreases.

In real-time compensation, both metrics tend to be appropriate when utilizing the linear approximation model. The best position error rate is 24.94%, and the distance error rate is 11.42%. However, as shown in Tables 3.10 and 3.11, suitable error model differ by route and the different is large. As shown in Fig. 3.12, the model using similarity transformation compensates the curved estimated trajectory to be excessive route distance. Therefore, the compensation algorithm for real-time compensation should be improved.

Currently, the similarity transformation model uses the newest sparse location only when the parameter describing this model updates. Therefore, it is possible that the parameter describing this model changes considerably every time this model updates.

Table 3.10 Distance error rates average of individual routes of real-time compensation utilizing the model using similarity transformation

	Without model (%)	Uniform (%)	One-side (%)	Unbalanced (%)
Route B	5.53	7.03	5.61	6.59
Route C	30.27	34.23	31.16	35.28

Table 3.11 Distance error rates average of indivisual routes of real-time compensation utilizing the model using linear approximation

	Without model (%)	Uniform (%)	One-side (%)	Unbalanced (%)
Route B	14.97	17.94	16.72	16.88
Route C	7.68	7.23	8.29	7.84

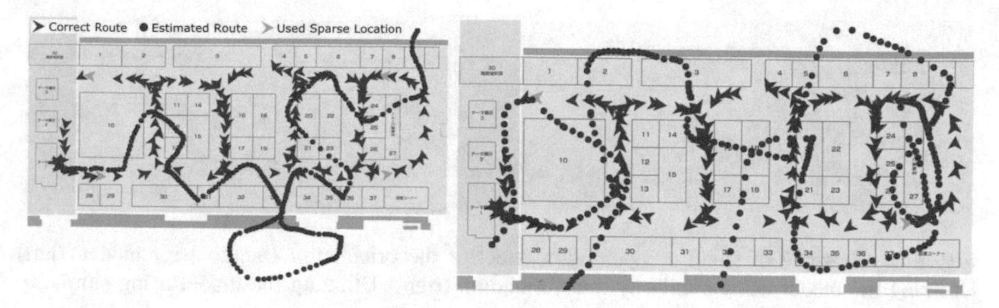

Fig. 3.12 Bad example of compensation effect. (left) No compensation. (right) Utilizing the model using similarity transformation (post-compensation)

Table 3.12 Position error rates average of individual routes utilizing the model using linear approximation

	Without model (%)	Uniform (%)	One-side (%)	Unbalanced (%)
Route A	14.01	13.26	15.40	11.25
Route B	17.66	7.58	8.61	7.06
Route C	12.43	11.44	8.90	9.16
Route D	23.94	29.82	25.62	26.50

3.4.4 Discussion About Error Model of Orientation

As shown in Table 3.12, the compensation effect is considerably different by route. There are cases in which the compensation effect of the orientation change error model is large or small, or it can even cause the model to function worse. The best improvement rate is 3.53% in route A, and 10.6% in route B. The compensated trajectory of route A is shown in Fig. 3.13, and the that of route B is shown in Fig. 3.14. It is considered that the quantity of the compensation effect is depend on the complexity of route. The harmful effect at the route that the compensation cause estimation worse is slite. Therefore, the advantage of the goodly compensation effect in the complex routes is superior to slitely harmful effect in the simple routes.

In this experiment, we cannot confirm which orientation change error model is superior. Conducting experiments with various routes and subjects, and further consideration will be needed to yield further findings.

Additionally, the compensation effect of the drift angle error model is imperceptible in this experiment. As future work, we need to evaluate this model by conducting experiments with a long straight route.

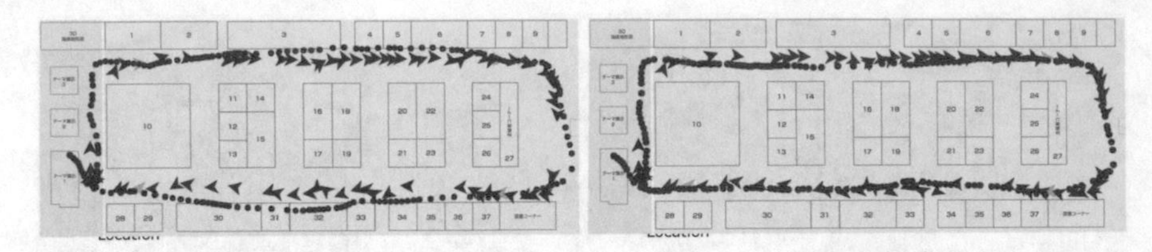

Fig. 3.13 Example of small compensation effect of the orientation change error model. (left) Utilizing the model using similarity transformation. (right) Utilizing the model using similarity transformation and the unbalanced error model

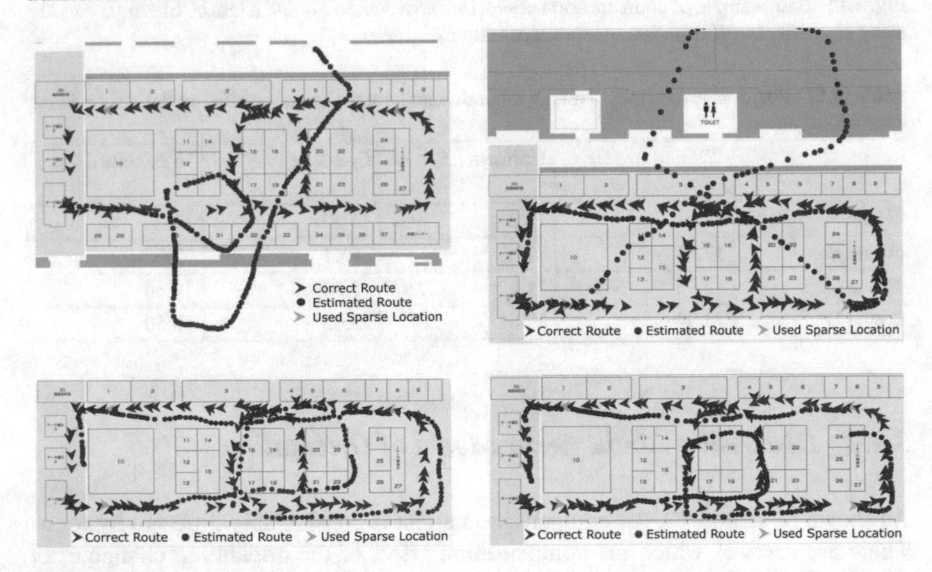

Fig. 3.14 Examples of successful compensations. (upper left) No compensation. (upper right) Utilizing the model using similarity transformation. (lower left) Utilizing the model using similarity transformation and the unbalanced error model. (lower right) Utilizing the model using linear approximation and the unbalanced error model

3.5 Conclusion

In this chapter, we proposed a compensation scheme using sparse locations and error models. We modeled errors that occurred by individual differences of step length or walking motion and updated the parameters describing the error models utilizing sparse locations. Then, we applied these models to PDR estimation. For evaluating our scheme, we collected the walking data of 2 subjects on four routes. As a result, we gained knowledge that an appropriate moving distance error model is different than post-compensation and real-time compensation. Moreover, using our scheme improved the position evaluation metric to approximately 10% and the distance evaluation metric to approximately 7%. Future work will address the following points:

- Improve the error models. In this research, current error models are linear. A nonlinear model may be designed. Additionally, we should evaluate the error model that has not been evaluated and improve the model.
- Use other devices to obtain sparse locations. Sparse locations that we are using currently are obtained by a UWB localization system. In practice, we should use sparse locations obtained from inexpensive devices, such as BLE. Therefore, a scheme considering estimation error of sparse locations is necessary.

Additionally, some existing PDR methods proposed better step detection and step length estimation algorithm. Improving PDR estimation without compensation, we should apply these methods to our PDR algorithm. Then, further evaluations in the other environment are required. For example, HASC-IPSC (Kaji et al. 2013) contains 3D floor plan of the target environment.

Acknowledgements Part of this research was supported by the Executive Committee of Geospatial EXPO 2016 indoor localization x IoT demonstration experiment, JSPS KAKENHI Grant Number JP 17H01762.

References

Abe M, Kaji K, Hiroi K, Kawaguchi N (2016) PIEM: path independent evaluation metric for relative localization. In: 2016 international conference on indoor positioning and indoor navigation (IPIN), pp 1–8

Ban R, Kaji K, Hiroi K, Kawaguchi N (2015) Indoor positioning method integrating pedestrian dead reckoning with magnetic field and WiFi fingerprints. In: 8th international conference on mobile computing and ubiquitous networking (ICMU). IEEE, pp 167–172

Chang Q, Van de Velde S, Wang W, Li Q, Hou H, Heidi S (2015) Wi-Fi fingerprint positioning updated by pedestrian dead reckoning for mobile phone indoor localization, pp 729–739

Ciabattoni L, Foresi G, Monteriù A, Pepa L, Pagnotta DP, Spalazzi L, Verdini F (2017) Real time indoor localization integrating a model based pedestrian dead reckoning on smartphone and BLE beacons. J Ambient Intell Hum Comput

Faragher R, Harle R (2015) Location fingerprinting with bluetooth low energy beacons. IEEE J Sel Areas Commun 33(11):2418–2428

Farshad A, Li J, Marina MK, Garcia FJ (2013) A microscopic look at WiFi fingerprinting for indoor mobile phone localization in diverse environments. In: International conference on indoor positioning and indoor navigation, vol 28, p 31

Ferris B, Fox D, Lawrence N (2007) WiFi-SLAM using Gaussian process latent variable models. In: 20th international joint conference on artificial intelligence, IJCAI'07. Morgan Kaufmann Publishers Inc., pp 2480–2485

Geospatial EXPO 2016. http://g-expo.jp/2016/. Accessed 01 Feb 2019

Kaji K, Watanabe H, Ban R, Kawaguchi N (2013) HASC-IPSC: indoor pedestrian sensing corpus with a balance of gender and age for indoor positioning and floor-plan generation researches. In: Proceedings of the 2013 ACM conference on pervasive and ubiquitous computing adjunct publication. ACM, pp 605–610

Kang W, Han Y (2015) SmartPDR: smartphone-based pedestrian dead reckoning for indoor localization. IEEE Sens J 15(5):2906–2916

Krach B, Roberston P (2008) Cascaded estimation architecture for integration of foot-mounted inertial sensors. In: 2008 IEEE/ION position, location and navigation symposium. IEEE, pp 112–119

Li J, Wang Q, Liu X, Cao S, Liu F (2014) A pedestrian dead reckoning system integrating low-cost MEMS inertial sensors and GPS receiver. J Eng Sci Technol Rev 7(2)
Miyake T, Arai I (2013) An adaptive step length reasoning for time periods and members of a user's party [in Japanese]. Spec Interes Group Tech Rep IPSJ 32:1–7
Nozaki J, Hiroi K, Kaji K, Kawaguchi N (2017) Compensation scheme for PDR using sparse location and error model. In: Proceedings of the 2017 ACM international joint conference on pervasive and ubiquitous computing and proceedings of the 2017 ACM international symposium on wearable computers, UbiComp '17, New York, NY, USA. ACM, pp 587–596
Oberli C, Torres-Torriti M, Landau D (2010) Performance evaluation of UHF RFID technologies for real-time passenger recognition in intelligent public transportation systems. IEEE Trans Intell Transp Syst 11(3):748–753
Shin B, Lee JH, Lee H, Kim E, Kim J, Lee S, Cho Y, Park S, Lee T (2012) Indoor 3D pedestrian tracking algorithm based on PDR using smarthphone. In: 12th international conference on control, automation and systems, pp 1442–1445
Takeshima C, Kaji K, Hiroi K, Kawaguchi N, Kamiyama T, Ohta K, Inamura H (2015) A pedestrian passage detection method by using spinning magnets on corridors. In: Adjunct proceedings of the 2015 ACM international joint conference on pervasive and ubiquitous computing and proceedings of the 2015 ACM international symposium on wearable computers. ACM, pp 411–414
Zheng L, Zhou W, Tang W, Zheng X, Peng A, Zheng H (2016) A 3D indoor positioning system based on low-cost MEMS sensors. Simul Model Pract Theory 65:45–56

Chapter 4
Towards the Design and Evaluation of Robust Audio-Sensing Systems

Akhil Mathur, Anton Isopoussu, Fahim Kawsar, Robert Smith, Nadia Berthouze and Nicholas D. Lane

Abstract As sensor-based inference models move out of laboratories into the real-world, it is of crucial importance that these models retain their performance under changing hardware and environment conditions that are expected to occur in-the-wild. This chapter motivates this challenging research problem in the context of audio sensing models, by presenting three empirical studies which evaluate the impact of hardware and environment variabilities on cloud-scale as well as embedded-scale audio models. Our results show that even the state-of-the-art deep learning models show significant performance degradation in the presence of ambient acoustic noise, and more surprisingly under scenarios of microphone variability, with accuracy losses as high as 15% in some scenarios. Further, we provide intuition on how this problem of model robustness relates to the broader topic of dataset-shift in the machine learning literature, and highlight future research directions for the mobile sensing community which include the investigation of domain adaptation and domain generalization solutions in the context of sensing systems.

4.1 Introduction

In recent years, we have witnessed a rapid increase in consumer devices and mobile sensing applications which aim to infer user context, activities and behavior from a variety of sensor data collected from the users. A number of commercial smart devices have already been launched in the market for monitoring a user's sleep

---

A. Mathur (✉)
Nokia Bell Labs and University College London, London, England
e-mail: akhil.mathur@nokia-bell-labs.com

A. Isopoussu · F. Kawsar
Nokia Bell Labs, London, England

R. Smith · N. Berthouze
University College London, London, England

N. D. Lane
University of Oxford, Oxford, England

N. Kawaguchi et al. (eds.), *Human Activity Sensing*,
Springer Series in Adaptive Environments,
https://doi.org/10.1007/978-3-030-13001-5_4

(Nokia sleep tracker 2018), physical activity (FitBit 2017), dietary actions (New technology tracks food intake by monitoring wrist movements 2017), stress (Empatica wristband 2018), daily activities (FitBit 2017), ambient environments (Narrative Clip 2017; Google Glass 2016), and emotional well-being (Empatica wristband 2018). In addition to these dedicated consumer devices, a range of mobile sensing applications have been proposed to detect context and activities such as sleep (Hao et al. 2013), exercise (Lu 2019) and transportation mode (Blunck 2013). In particular, owing to the recent breakthroughs in machine learning techniques for audio-processing, a number of promising audio sensing systems and applications have been proposed, including those which infer a user's emotion (Rachuri 2010), eating episodes (e.g., chewing) (Amft et al. 2005), and speech characteristics (e.g., speaker verification) (Variani et al. 2014), keyword spotting (Chen et al. 2014). The advancements in audio-based inference models are also ushering in the design of open audio-based hardware platforms which allow developers to create powerful audio services for the end-users. For instance, by integrating off-the-shelf microphone arrays with embedded platforms such as Raspberry Pi and cloud-based audio sensing models, developers can rapidly create their own version (Hardware to emulate amazon echo 2019) of a speech processing device such as an Amazon Echo.

As sensory inference systems move out of the laboratory setting into the wild, it is imperative that they work robustly on thousands and millions of end-user devices in unconstrained real-world scenarios. Indeed, prior research has highlighted it as a major research challenge to make sensory systems robust against hardware, software, environment and user variabilities. Blunck et al. (2013) demonstrated how GPS sensor variability can impact the data quality and the performance of inference models on smartphones. Stisen et al. (2015) studied sampling rate heterogeneity in inertial sensors of smart devices and found that software factors such as instantaneous CPU loads can cause a large variability in the accelerometer outputs of smartphones and smartwatches. Chon et al. (2013) found that sound classification models show poor accuracies when deployed in unconstrained environments. Similar findings were shown by Lee et al. (2013) about the adverse impact of acoustic environments on speaker turn-taking detection. Vision models are also impacted by environmental variabilities such as lighting conditions (Yang et al. 2016), various forms of object occlusion (Chandler and Mingolla 2016), and operation variabilities such as blurry, out-of-focus images due to unstable cameras.

In this chapter, we focus our attention on the robustness of audio-sensing models in unconstrained real-world scenarios. While a number of factors can influence the performance of an acoustic inference model in practice, this chapter explores two forms of prominent noise that these models are expected to encounter in the real-world:

Acoustic Environment Noise: An audio-sensing application should ideally make accurate inferences irrespective of where and when it is used. However in practice, the environment (e.g., cafe, train station) and environmental conditions (e.g., raining, ambient music) in which an audio signal is captured add background noises to the signal that may confuse the underlying inference models and impact their accuracy.

As such, one of the desired properties for audio-based inference systems is their robustness to diverse acoustic environments.

Microphone Heterogeneity: Audio inference models, once trained, are deployed on numerous mobile and wearable devices, many of which are not known while the models are trained and could come from different hardware manufacturers. This is a challenging scenario because different manufacturers use different hardware components (i.e., microphones) and may also have different software pipelines which process the raw audio signal before exposing them to user applications. Therefore, inference models need to be robust against these forms of microphone heterogeneity expected in the wild.

In this chapter, we build upon our prior work (Mathur et al. 2018) and study the performance of two widely-used general-purpose audio models under scenarios of real-world noise. First, we study how microphone heterogeneity impact cloud-based automatic speech recognition (ASR) models and thereafter, we extend this analysis to a small footprint keyword detection model. For our experiments, we use off-the-shelf microphones ranging from mid-range microphone arrays to low-cost USB microphones. In addition to the microphone heterogeneity problem, we also study the impact of background noise on cloud-based ASR models.

Quite unexpectedly, we find significant difference in the performance of our target audio models when they are exposed to different microphones, with observed accuracy drops as high as 15% when models trained on one microphone are deployed on another. Our results also reveal that cloud-based ASR models are more tolerant to ambient acoustic noise and show reasonable performance under moderate amounts of ambient noise, however the error rates increase significantly as the noise power is increased. We conclude the chapter by introducing the general topic of domain adaptation in machine learning and ways of leveraging domain adaptation techniques for improving the robustness of sensing models. More specifically, using microphone heterogeneity as a use-case, we discuss techniques which could be used at training-time and inference-time—depending on the system requirements—to improve the robustness of audio models when the training and test microphone differ. Taken together, our analysis and findings suggest the need for more rigorous evaluation of sensor-based inference systems, going beyond the conventional evaluation techniques such as train-test split and cross-validation.

4.2 Methodology

We now discuss our methodology for evaluating the robustness of audio models in real-world scenarios.

Audio Tasks and Datasets: Two representative audio tasks and datasets are used in our analysis:

- Automatic Speech Recognition (ASR): ASR is a fundamental component of audio- or speech-processing systems and recent advances in the field of deep learning have significantly improved the performance of ASR models (Hannun et al. 2014). Our experiments are conducted on the Librispeech-clean (Panayotov et al. 2015) dataset, which is a widely-used ASR benchmark dataset for comparing the accuracy of different ASR models. We use 1000 randomly selected test audios from the Librispeech-clean dataset, with an average duration of 7.95 s and sampling rate of 16,000 Hz. In the rest of the chapter, we refer to this dataset as *Librispeech-clean-1000*.

- Keyword Detection: In this task, the goal is to identify the presence of a certain keyword class (e.g., Hey Alexa) in a given speech segment. We use the *Speech Commands* dataset containing 65,000 one-second long utterances of 30 short keywords (Speech Commands Dataset 2018) for our experiments. Instead of using all 30 classes, we used a subset of 12 classes (yes, no, up, down, left, right, on, off, stop, go, zero, one) for our analysis.

Audio Models: We now describe the two audio models on which the above mentioned datasets were evaluated:

- ASR Models: We conduct our experiments on ASR models from Google (using the Google Speech API 2019) and Microsoft (using the Bing Speech API 2019). The models use a CNN-bidirectional LSTM model structure (Xiong et al. 2017) and have shown near-human accuracy on ASR tasks (Microsoft speech recognition 2019; Google speech recognition 2019). Audios from the *Librispeech-clean-1000* dataset under both experiment conditions were passed to the models through REST APIs, and Word Error Rate (WER) was computed on the ASR transcripts.
- Keyword Detection Model: We use a small-footprint keyword detection architecture proposed in Zhang et al. (2017) to train the model. The input to this model is a two-dimensional tensor extracted from the one-second long keyword recording, consisting of time frames on one axis and 24 MFCC features on the other axis. The model outputs a probability of a given audio recording belonging to a certain keyword class (e.g., Yes, No) or to an Unknown class.

Experiment Conditions: As discussed earlier, our investigation of audio model robustness focuses on two key sources of noise observed in audio signals in real-world scenarios:

- Microphone Heterogeneity: To evaluate how audio models cope against microphone variability, we needed to record a large-scale test dataset from different microphones under the same environment conditions. For this, we replayed the *Librispeech-clean-1000* and *Speech Commands* datasets on a JBL LSR 305 monitor speaker[1] and recorded them simultaneously on three different microphones namely Matrix Voice (2019), ReSpeaker (2019) and PlugUSB in a quiet environment. While the first two microphones are multi-channel microphone arrays commonly used in consumer devices such as Amazon Echo, the last microphone is a low-cost USB microphone compatible with embedded platforms such as Raspberry Pi. The microphones were kept at a distance of 10 cm from the speaker in

[1] We chose this speaker due to its flat frequency response in the human speech frequency range.

order to minimize the effect of room acoustics on the recorded audio. In effect, we created four variants each of the *Librispeech-clean-1000* and *Speech Commands* datasets, including the original dataset and the three re-recordings that we did with off-the-shelf embedded microphones.

- Acoustic Environment noise: To simulate the effect of different acoustic environments, we mix the speech audios from Librispeech dataset with examples of real-world background noise taken from the ESC-50 dataset (Piczak 2015). To this end, we randomly sampled 200 audios from the *Librispeech-1000* dataset and augmented them with background audios of *Rain* and *Wind* from the ESC-50 dataset.

4.3 Results

Figures 4.1 and 4.2 show the effect of microphone variability on the accuracy of the ASR models. Firstly, we observe that for all three microphones, the word error rate (WER) increases over the baseline (i.e., the original Librispeech audios) by as high as 1.41 times. More importantly, the model performance varies across different microphones (e.g., from 1.24x to 1.41x WER increase in the case of Bing ASR model), which suggests that the ASR models are not completely robust to microphone variability. Similar trends are observed with the keyword detection model. Figure 4.3 shows that when the training and test devices are the same, the keyword detection model provides the highest accuracy. However, when there is a mismatch between the training and test devices, it causes a significant degradation in accuracy as high as 15%.

Further, in Fig. 4.4, we plot the spectrograms of an audio segment from the Librispeech-1000 dataset in its original form Fig. 4.4a as well as when it is captured by different microphones Fig. 4.4b–d. Subtle variabilities in how different microphones capture the same audio signal can be observed from the figures, and we hypothesize that the ASR models are not trained to account for these variabilities, which in turn leads to varying levels of increase in the WER.

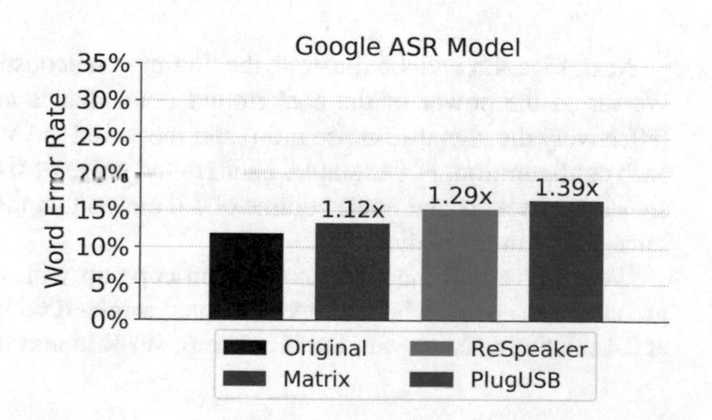

Fig. 4.1 Impact of the microphone variability on Google ASR model. Values on the bars illustrate the increase in WER over the original audio WER (black bar)

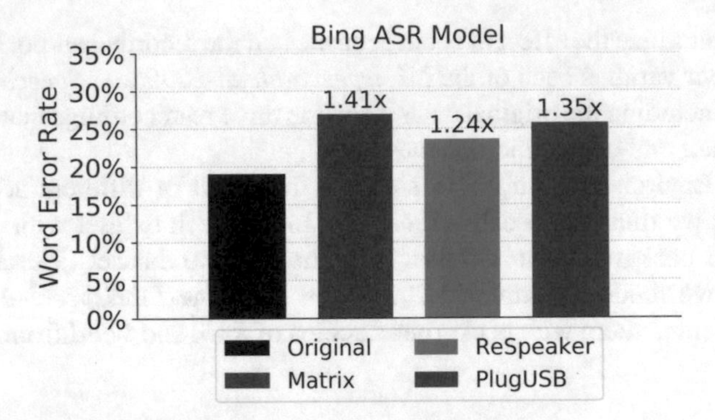

Fig. 4.2 Impact of microphone variability on Bing ASR model. Values on the bars illustrate the increase in WER over the original audio WER (black bar)

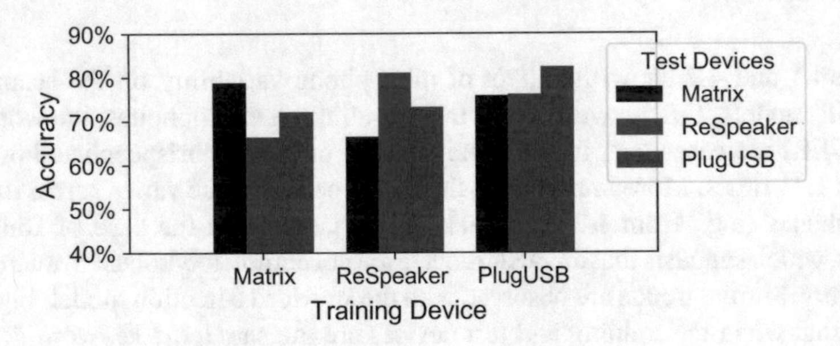

Fig. 4.3 Impact of microphone variability on the keyword detection model

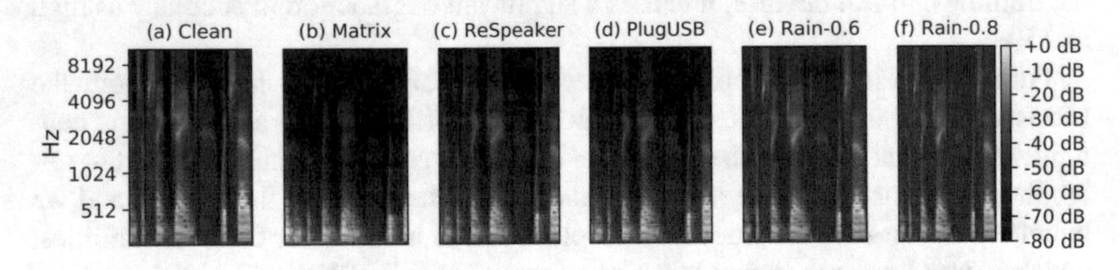

Fig. 4.4 Mel-Scale spectrograms of an audio segment under different experiment conditions

Next, Figs. 4.5 and 4.6 illustrate the findings on acoustic environment robustness. We varied the power of the background noise that is added to the speech signal (effectively the signal-to-noise ratio) and measured the WER of the ASR models in each configuration. For example, background noise of 0.0 corresponds to the clean signal and background noise volume of 1.0 means that the signal and noise have the same power in the audio.

We observe that the ASR models can cope up with moderate amount of background noise—e.g., when the speech signal is mixed with 'Wind' and 'Rain' audios at 0.4 relative noise power, the increase in WER is less than 1.25x for both Google

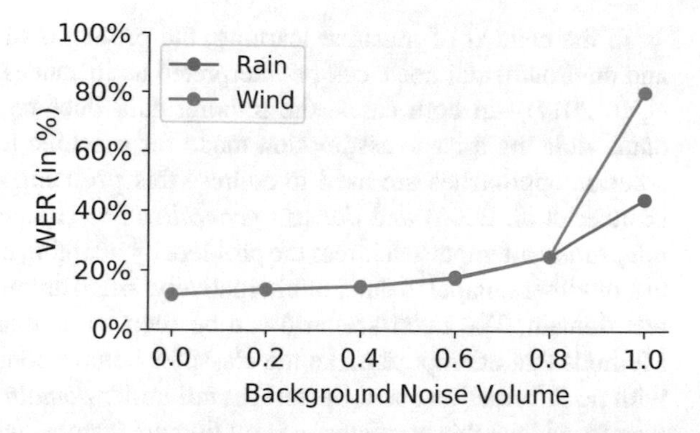

Fig. 4.5 Effect of two types of background noise on Google ASR model

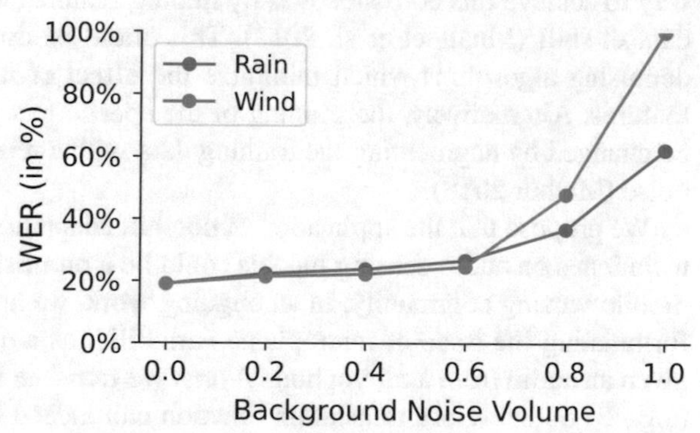

Fig. 4.6 Effect of two types of background noise on Bing ASR model

and Bing ASR models. However, when the relative noise power is increased to 0.8, the WER increases by more than 2x above the baseline for both the models.

Finally, we make the following observation on the comparative robustness of the ASR models to microphone variability and environment noise. In Fig. 4.4, although the Rain-0.6 spectrogram (Fig. 4.4e) looks visibly more noisy than the spectrograms collected from different microphones (Fig. 4.4b–d), the performance of ASR models on Rain-0.6 dataset is similar to that on various microphones. This indicates that the ASR models are able to cope with background noise in the speech much better than the subtle variabilities caused by different microphones. Further research is needed to uncover the underlying causes behind this behavior.

4.4 Discussion and Future Directions

Our experiments show that deep learning based audio models are not robust to real-world noise caused by microphone variability and different acoustic environments. In this section, we broadly discuss the research directions that could be explored to solve this problem.

In the context of machine learning, the problems of microphone heterogeneity and environmental noise can be interpreted as instances of *dataset shift* (Sugiyama et al. 2017)—in both cases, the training data does not accurately reflect the test data, violating a basic assumption made for machine learning models. Two broad solution approaches are used to address this problem, namely *domain adaptation* (Blitzer et al. 2006) and *domain generalisation* (Blanchard et al. 2011). *Domain adaptation* attempts to address the problem by adapting an existing model by making use of either unlabeled data, or alternatively, small amounts of labeled data from the test domain. The latter scenario can be seen as an example of *transfer learning*. Methods that attempt to make the classifier behave consistently under dataset shift with no information about the test set fall under *domain generalization*. The easiest way to achieve this consistency is by finding features which are invariant under the dataset shift (Muandet et al. 2013). This could be done by designing specialized denoising algorithms which minimize the effect of noise sources on the learned features. Alternatively, the training of the speech recognition algorithm may itself be changed by augmenting the training data with a representative range of types of noise (Mathur 2018).

We propose that the application of domain adaptation and domain generalization techniques on audio-sensing models could be a promising research direction for the mobile sensing community. In an ongoing work, we are exploring the feasibility of formulating the issue of microphone variability as a data translation problem, i.e., given an audio from a microphone A, can we translate it to a different microphone's (e.g., B) domain? If a translation function can indeed be learned between a pair of microphones, it can subsequently be used to convert any audio training data across microphone domains, and audio models could be trained on such diverse training datasets. One key challenge however is that generating large aligned audio datasets to train the translation models discussed above can be hard. Each time a new microphone is considered, in order to produce an aligned version of the dataset, the entire dataset needs to be recorded using the new microphone. This leads to major issues in scaling the approach to multiple devices.

As such, we are exploring solutions which can learn the mapping between two microphones without requiring time-aligned data from them. To this end, we propose to use the CycleGAN architecture introduced in Zhu et al. (2017), which involves simultaneously training two translation models, one mapping the training domain to the test domain, and another one in reverse. The model also uses a cycle loss as a way to improve the performance of both translation models. Once trained, our translation model named Mic2Mic can be used in two different ways.

- Training time: As shown in Fig. 4.7, Mic2Mic can be used to translate the entire training dataset from microphone A to microphone B, as a way to augment the training data to add awareness about the properties of the test microphones into the audio model training process. The original training dataset is then combined with the translated dataset to generate an augmented dataset, upon which the task-specific audio model is trained.

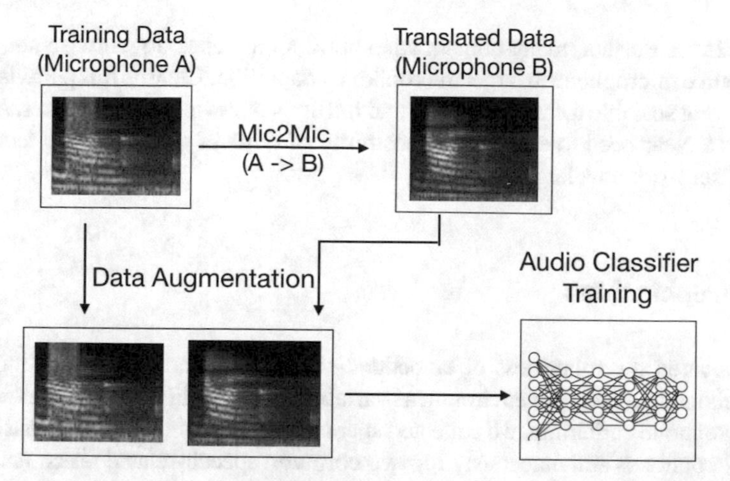

Fig. 4.7 Translation model deployed as a data augmentation to correct for microphone variability

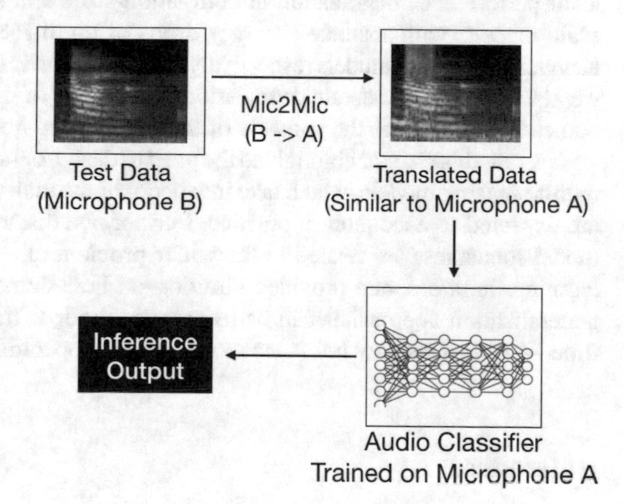

Fig. 4.8 Translation model deployed to translate test time data to the training data distribution to correct for microphone variability

- Inference time: Alternatively, Mic2Mic can be deployed in the inference pipeline of audio models as a real-time translation component. As shown in Fig. 4.8, audio data from the test microphone at inference-time is first passed to Mic2Mic, which transforms it to bring it closer to the training data distribution. Thereafter, the translated data is inputted to the pre-trained audio classifier to obtain the inferences.

While our initial results are promising, there remain a number of open research challenges to make Mic2Mic and other domain adaptation challenges work in uncontrolled settings. In real-world settings, it is likely to encounter a combination of different variabilities in the sensor data. For instance, microphone variability can combine with acoustic environmental noise and user-specific behavior, and will need a much more complex solution than the single-variability adaptation approaches such as Mic2Mic. Further, many domain adaptation approaches do pairwise adaptation, e.g., source microphone to target microphone adaptation. Clearly, this pairwise adaptation is not scalable for the thousands and millions of devices in the market. As such, there is a clear need to explore the feasibility of domain generalization techniques which can work on a larger scale.

4.5 Conclusions

We evaluated the robustness of embedded-scale and cloud-scale audio models to microphone and acoustic environment variability. To facilitate our first experiment on microphone variability, we collected speech samples from three different embedded microphones simultaneously for two common speech-related tasks: Automatic Speech Recognition (ASR) and Keyword Detection. Our results demonstrate significant performance degradation in both cloud-scale and small footprint embedded-scale models, with absolute accuracy drops of up to 7% and 15% in the ASR and Keyword Detection models respectively. For the acoustic background noise scenario, we also observe a moderate degradation in accuracy of the audio models, which becomes more severe as the intensity of the background noise increases.

Overall, this chapter highlighted the need to design better evaluation techniques for mobile sensing models, which take into account the real-world noise that the models are expected to encounter in practice. Further, we discussed that the challenges of model robustness are related to the wider problem of dataset-shift in the machine learning literature, and provided intuition on how domain adaptation and domain generalization approaches can be leveraged—both at training-time and inference-time—to adapt sensory inference models to new operating scenarios.

References

Nokia sleep tracker (2018). https://health.nokia.com/uk/en/sleep/
FitBit (2017). https://www.fitbit.com
New technology tracks food intake by monitoring wrist movements. http://gadgetsandwearables.com/2017/03/29/food-tracking/ (2017). Accessed 20 June 2019 10:48:03
Empatica wristband (2018). https://www.empatica.com/en-gb/research/e4/. Accessed 20 June 2019 10:48:03
Narrative clip (2017). http://getnarrative.com/narrative-clip-1. Accessed 1 Sept 2017
Google glass (2016). https://developers.google.com/glass/distribute/glass-at-work. Accessed 20 June 2019 10:48:03
Hao T, Xing G, Zhou G (ACM, 2013), SenSys'13. https://doi.org/10.1145/2517351.2517359
Lu HEA (2019) Proceedings of Sensys '10. ACM, pp 71–84
Blunck H et al (2013) In: Proceedings of the 2013 ACM Ubicomp. ACM, pp 1087–1098
Rachuri K et al (2010) In: Proceedings of Ubicomp'10. ACM, pp 281–290
Amft O, Stäger M, Lukowicz P, Tröster G (2005) In: Ubicomp Springer, pp 56–72

Variani E, Lei X, McDermott E, Moreno IL, Gonzalez-Dominguez J (2014) In: 2014 IEEE International conference on acoustics, speech and signal processing (ICASSP). IEEE, pp 4052–4056
Chen G, Parada C, Heigold G (2014) In: ICASSP. IEEE, pp 4087–4091
Hardware to emulate amazon echo. https://tinyurl.com/y84d6r2n/
Stisen A et al (2015) In: Proceedings of Sensys. ACM, pp 127–140
Chon Y et al (2013) In: Ubicomp. ACM, pp 3–12
Lee Y, Min C, Hwang J, Lee I, Hwang Y, Ju C, Yoo M, Moon U, Lee J, Song J (2013) In: Proceeding of Mobisys'13. ACM, pp 375–388
Yang S, Wiliem A, Lovell BC (2016) In: 2016 international conference on image and vision computing New Zealand (IVCNZ). IEEE, pp 1–6
Chandler B, Mingolla E (2016) Computational intelligence and neuroscience
Mathur A, Isopoussu A, Kawsar E, Smith R, Lane ND, Berthouze N (2018) In: Proceedings of the 2018 ACM international joint conference and 2018 international symposium on pervasive and ubiquitous computing and wearable computers. ACM, New York, NY, USA, 2018) UbiComp'18, pp 1409–1413. https://doi.org/10.1145/3267305.3267505
Hannun A, Case C, Casper J, Catanzaro B, Diamos G, Elsen E, Prenger R, Satheesh S, Sengupta S, Coates A, et al (2014) arXiv:1412.5567
Panayotov V, Chen G, Povey D, Khudanpur S (2015) In: ICASSP. IEEE, pp 5206–5210
Speech Commands Dataset. https://research.googleblog.com/2017/08/launching-speech-commands-dataset.html. Accessed 20 June 2019 10:48:03
Google Speech API. https://cloud.google.com/speech-to-text/
Bing Speech API. https://azure.microsoft.com/en-us/services/cognitive-services/speech/
Xiong W, Wu L, Alleva F, Droppo J, Huang X, Stolcke A (2017) ArXiv e-prints
Microsoft speech recognition. https://tinyurl.com/ybnm9zdj/
Google speech recognition. https://tinyurl.com/y7dm37vw/
Zhang Y, Suda N, Lai L, Chandra V (2017) arXiv:1711.07128
Matrix Voice. https://www.matrix.one/products/voice/
ReSpeaker. https://respeaker.io/
Piczak KJ (2015) In: ACM multimedia. ACM, pp 1015–1018
Sugiyama M, Lawrence ND, Schwaighofer A et al (2017) Dataset shift in machine learning. The MIT Press
Blitzer J, McDonald R, Pereira (2006) In: Proceedings of the 2006 conference on empirical methods in natural language processing, pp 120–128
Blanchard G, Lee G, Scott C (2011) Advances in neural information processing systems 2178–2186
Muandet K, Balduzzi D, Schölkopf B (2013) ICML 10–18
Mathur A et al (2018) In: IPSN. IEEE
Zhu JY, Park T, Isola P, Efros AA (2017) CVPR 2223–2232

Chapter 5
A Wi-Fi Positioning Method Considering Radio Attenuation of Human Body

Shohei Harada, Kazuya Murao, Masahiro Mochizuki and Nobuhiko Nishio

Abstract The importance of location information is increasing more and more. Therefore, research on indoor positioning technology is actively conducted. Among them, Wi-Fi positioning is attracting attention because of its low introduction cost. However, since the Wi-Fi is blocked by the human body, the RSSI decreases. Therefore, we investigated radio wave attenuation by human body in preliminary experiments. We describe related research and its problems.

5.1 Introduction

5.1.1 Background and Purpose

Services using location information of users are increasing. In the outdoors, positioning is mainly performed using GPS (Global Positioning System). However. There is a problem that positioning can not be performed indoors because GPS radio waves can not be received due to obstacles such as walls and ceilings. For this reason, many studies have been conducted on methods for acquiring location information indoors. For example, PDR (Pedestrian Dead Reckoning) or Wi-Fi positioning using an acceleration sensor or a gyro sensor. Wi-Fi positioning has attracted attention because of

S. Harada (✉)
Graduated School of Information Science and Engineering, Ritsumeikan University, Kyoto, Shiga, Japan
e-mail: rama@ubi.cs.ritsumei.ac.jp

K. Murao · N. Nishio
College of Information Science and Engineering, Ritsumeikan University, Kyoto, Shiga, Japan
e-mail: murao@cs.ritsumei.ac.jp

N. Nishio
e-mail: nishio@is.ritsumei.ac.jp

M. Mochizuki
Research Organization of Science and Technology, Ritsumeikan University, Kyoto, Shiga, Japan
e-mail: moma@ubi.cs.ritsumei.ac.jp

N. Kawaguchi et al. (eds.), *Human Activity Sensing*,
Springer Series in Adaptive Environments,
https://doi.org/10.1007/978-3-030-13001-5_5

its ability to estimate absolute position and low introduction cost. Wi-Fi positioning mainly uses RSSI (Received Signal Strength Indicator). However, Wi-Fi is sensitive to the reflection and shielding of radio waves because it uses a communication band of 2.4 GHz band. Therefore, in Wi-Fi positioning, a drop in RSSI due to the human body blocking radio waves causes a drop in positioning accuracy. In this research, we try to improve the positioning accuracy in consideration of the attenuation of radio waves by the human body in Wi-Fi positioning. There are two methods in the proposed method, one is to create a radio environment map that is not influenced by the human body, and the other is to distinguish the base station where the radio attenuation has occurred. There are two contributions to this research. One is to identify the base station where the attenuation of radio waves by the human body occurred. The other is to improve the accuracy of Wi-Fi positioning by correcting the attenuation of radio waves.

5.1.2 Related Work

Murata et al. (2018) made positioning using BLE which is 2.4 GHz similar to Wi-Fi. When creating the radio wave map, radio waves were collected using an electric wheelchair. At that time, radio waves were collected in two directions in consideration of radio wave attenuation by the human body. However, there is a problem that the beacon arrangement is high density, the introduction cost is high, and it can not cope with the environment where Wi-Fi is installed on the ground by the colonnade. Kaji and Kawaguchi (2012) proposed a method for estimating the position by expressing the radio environment map with GMM (Gaussian Mixture Model) and applying a particle filter. GMM is a linear combination of a plurality of normal distributions, and the radio wave information of the base station is represented by this two-dimensional GMM. Each GMM normal distribution has an average μ and a variance covariance matrix Σ, and a mixing coefficient π_k as parameters, and these are obtained by EM algorithm. By modeling each base station in this way, a radio wave environment map for Wi-Fi positioning is created. Apply a particle filter to the generated radio wave map and radio wave information received at positioning to estimate the position. The particle filter is a method of predicting the next state by moving finite number of particles based on prediction and repeating likelihood calculation and resampling.

5.1.3 Preliminary Experiment

5.1.3.1 Investigation of Radio Wave Attenuation by Human Body

We conducted preliminary experiments to investigate radio attenuation by the human body. We conducted radio observations 100 times while changing the distance and direction from the Wi-Fi base station in our laboratory building. The distance from the base station ranges from 1 to 10 m in increments of 1 m, and the direction of the human body is two relative to the base station: the front side and the back side.

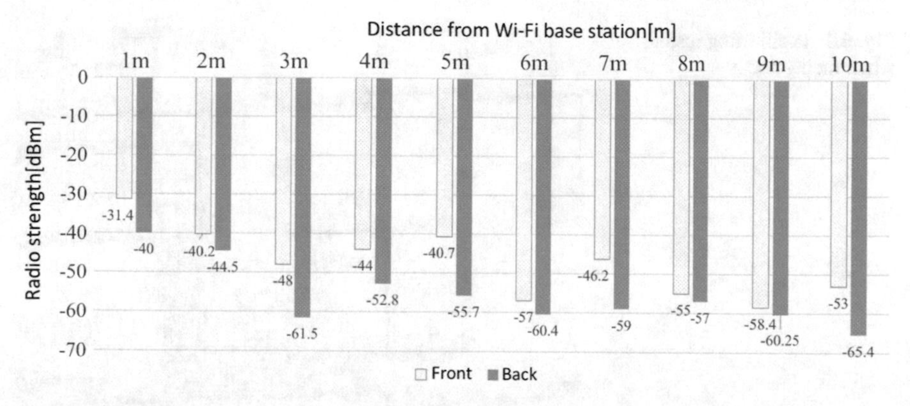

Fig. 5.1 RSSI attenuation survey result by human body

Experimental results are shown in Fig. 5.1. An average attenuation of 8.265 dBm occurred by the human body. The fact that the distance from the base station is not inversely proportional to RSSI is considered to be the effect of multipath fading.

5.1.3.2 Investigation of Influence of Radio Wave Attenuation by Human Body on Positioning Result

Preliminary experiments were conducted to investigate the effect of radio wave attenuation by the human body on the positioning result. In front of the elevator which is surrounded by the base station on the first floor of our laboratory building, Wi-Fi positioning was done about 100 times in the four directions of east, west, north and south at the same spot using NEXUS 5. The way of possession of NEXUS 5 is limited to hand held in front of the chest. The red marker in the figure is the positioning result, the blue marker is the average positioning result, and the green marker is the ground truth. The results in the northward direction are shown in the Fig. 5.2. The result in the eastward direction is shown in Fig. 5.3. The result in the south direction is shown in Fig. 5.4. The result of the westward orientation is shown in Fig. 5.5. It can be confirmed that the distribution of the positioning results is greatly different despite the fact that the position was estimated at the same point. It can be predicted that this is because base stations where radio wave attenuation by the human body occurs are different depending on the direction of the user.

5.2 Approach and Evaluation

In this study, radio wave model is approximated by GMM, and position estimation is performed using particle filter. We propose a method that does not cause radio attenuation by the human body during radio wave observation for radio wave model creation. Furthermore, we propose a method to identify the base station where radio

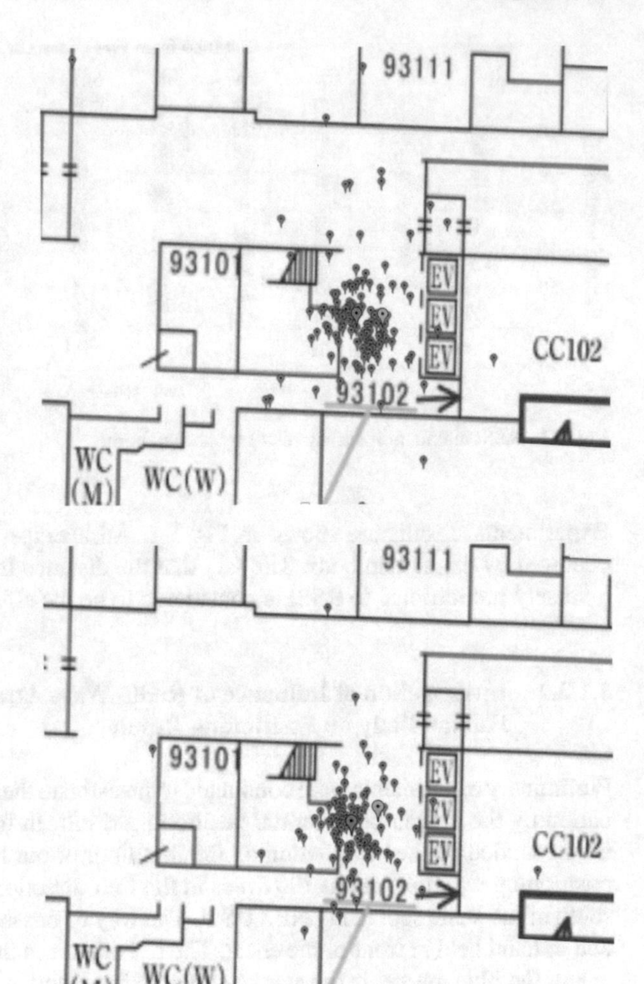

Fig. 5.2 Positioning result when facing north

Fig. 5.3 Positioning result when facing east

wave attenuation occurs by the human body. By using the proposed method, we can identify the base station where radio wave attenuation occurred and reduce the variance of the positioning result.

5.2.1 Proposed Method

5.2.1.1 Radio Environment Map Creation Not Affected by Radio Wave Attenuation

The radio wave environment map was created based on the method of Kaji and Kawaguchi (2012). However, radio wave attenuation by the human body can occur not only when positioning but also when creating a radio wave environment map. Therefore, using the self-shooting stick, the attitude of the observer was lowered, and

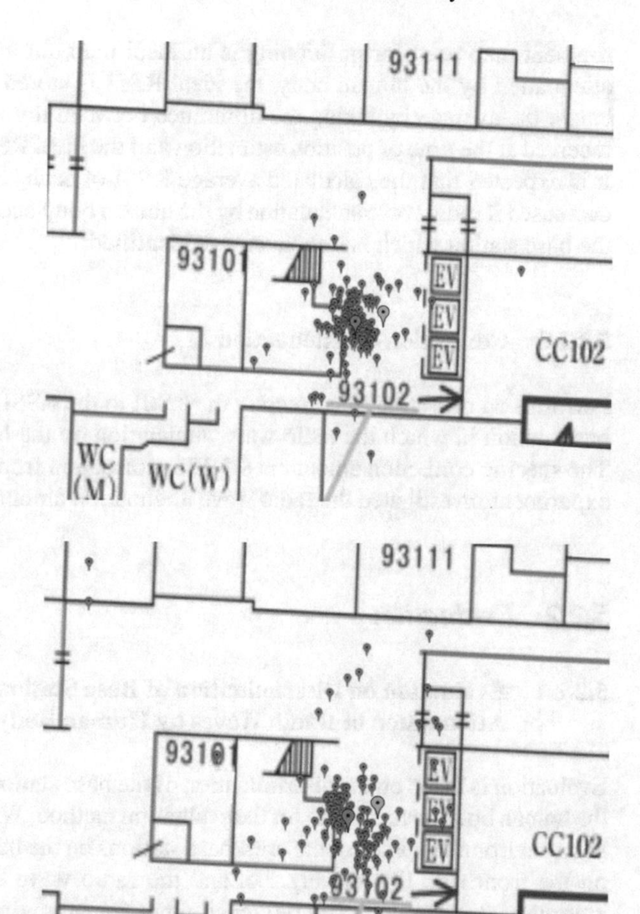

Fig. 5.4 Positioning result when facing south

Fig. 5.5 Positioning result when facing west

radio waves were collected so that radio wave attenuation by the human body does not occur at the height of about 140 cm which is the assumption that the terminal was hand held in front of the chest. In addition, because there is fluctuation in RSSI, the maximum value considered to be the least attenuated among radio wave collections of multiple times was used.

5.2.1.2 Algorithm for Discriminating the Attenuated Base Station

First, position estimation is performed using Wi-Fi radio waves including radio wave attenuation by the human body. After that, particles are uniformly dispersed to the radius of 10 m from the position estimation result. Here, since the radio wave environment map used for positioning is an ideal map not affected by the radio wave attenuation by the human body, the ideal RSSI is stored in each particle. We calculate the average by taking the difference between the RSSI of each base station received at the time of position estimation and the ideal RSSI stored in each particle. It is expected that the calculated average RSSI of each base station will be unduly decreased if radio wave attenuation by the human body occurs. By the above method, the base station which has attenuated is identified.

5.2.1.3 Correction of Attenuation

Performs an operation of correcting the RSSI to the RSSI before attenuation for the base station in which the radio wave attenuation by the human body has occurred. The specific correction amount is 8.3 dBm correction from the result of preliminary experiment investigated the radio wave attenuation amount of the human body.

5.2.2 Evaluation

5.2.2.1 Evaluation on Discrimination of Base Stations in Which Attenuation of Radio Waves by Human Body Occurred

Evaluation is made on the discrimination of the base station where the attenuation by the human body occurred. As for the evaluation method, Wi-Fi was observed 20 times in an environment where there are 4 base stations on the back side and 2 base stations on the front side like the Fig. 5.6, and the radio wave attenuation discrimination algorithm was applied. The reference point for arranging the particles was set to 10 m around the grand truth, and the average RSSI was used for each observation. Results are shown in Figs. 5.7, 5.8, 5.9 and 5.10 by number of observations. The black dashed line is the base station located in front of the user and the red dashed line is the base station located on the back of the user. The horizontal axis represents the difference from the ideal RSSI.The base station in front of the user is close to the ideal RSSI value. The base station behind is far from the ideal RSSI value. It was confirmed by the proposed method that effective values were calculated irrespective of the number of observations. In addition, we can also confirm that there are base stations that do not attenuate much even on the back side.

5.2.2.2 Evaluation on Positioning Result Correcting Radio Wave Attenuation by Human Body

Evaluation is performed on the positioning result correcting attenuation by the human body. Experimental method was stationary on the 1st floor of our research building west and stationed about 100 times. The experimental terminal used NEXUS 5.

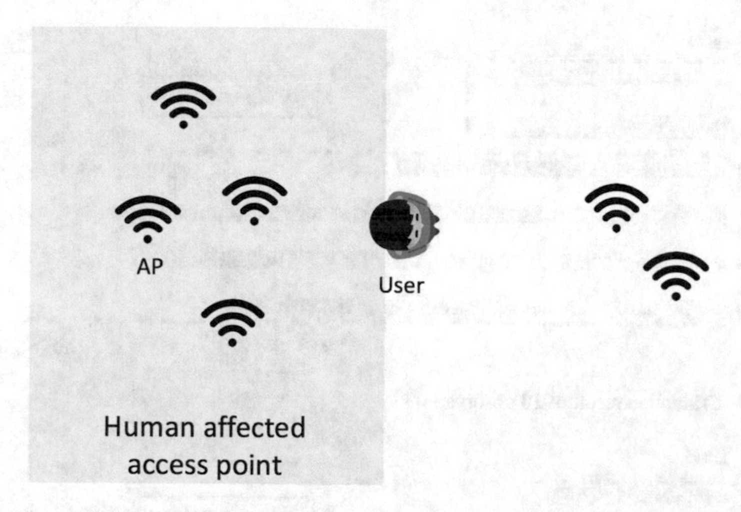

Fig. 5.6 The position relationship between AP and user for evaluation of determination of radio wave attenuation

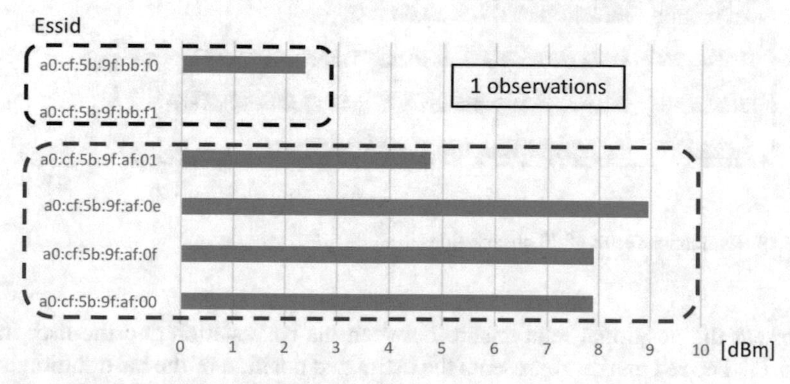

Fig. 5.7 Evaluation result of 1 observation

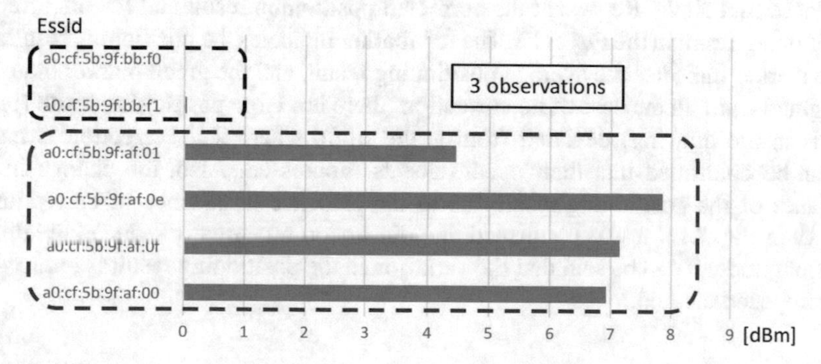

Fig. 5.8 Evaluation result of 3 observations

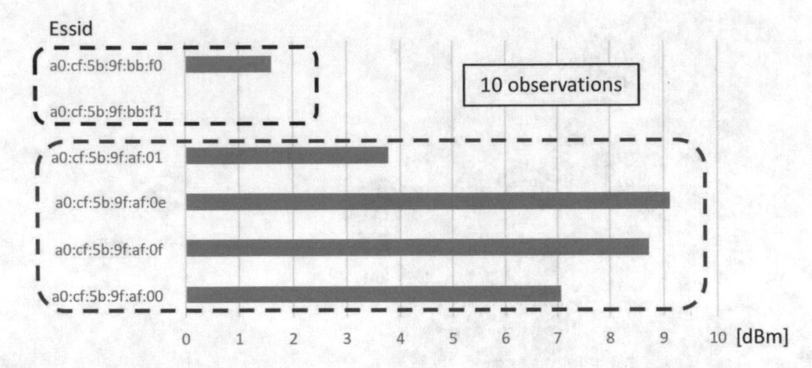

Fig. 5.9 Evaluation result of 10 observations

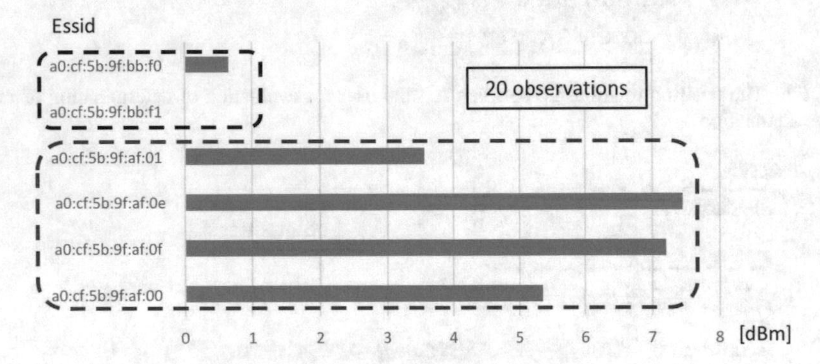

Fig. 5.10 Evaluation result of 20 observations

Represent the positional relationship between the base station and the user in the Fig. 5.11. The red marker represents the estimated position of the base station based on the radio wave environment map. We corrected the RSSI of the base station enclosed in a circle. Represent the corrected positioning result and the uncorrected positioning result in the Fig. 5.12. The red marker indicates the positioning result, the blue marker indicates the average positioning result, and the green marker indicates the grand truth. In the case of no correction, there is a large positioning result (black circle in the drawing) deviated from grand truth, whereas if correction is made, it can be confirmed that their occurrence is suppressed. Also, the change in the variance of the positioning result due to the presence or absence of correction is shown in Fig. 5.13. It was confirmed that dispersion was greatly reduced in latitude and longitude. It can be said that the variation in the positioning result is reduced by the proposed method.

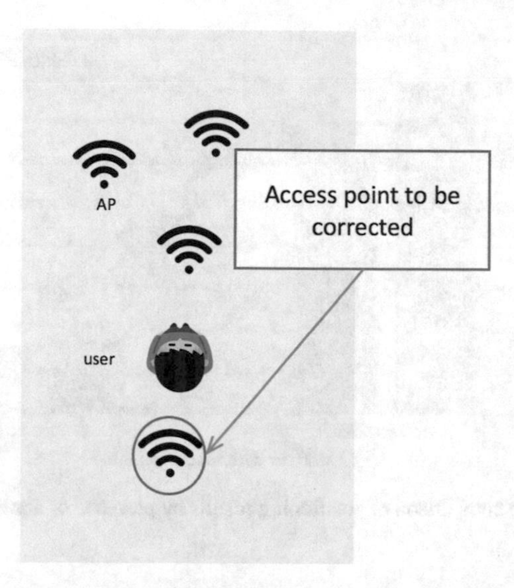

Fig. 5.11 The position relationship between AP and user for evaluation of positioning result based on presence/absence of correction

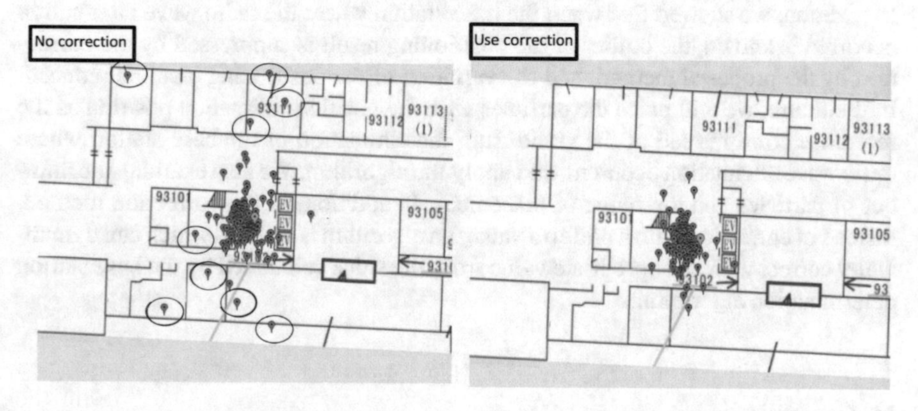

Fig. 5.12 Positioning result comparison based on presence/absence of correction

5.3 Conclusion

Initially, we discussed the problem that positioning error occurs due to radio wave attenuation by the user's body despite the fact that the existing Wi-Fi positioning method mainly uses RSSI. In preliminary experiments, we investigated and confirmed how much the radio wave attenuation by the human body and the radio wave attenuation by the human body affect the positioning result. Therefore, we proposed a method to create an ideal radio wave environment map with no radio wave attenuation and a method of discriminating/correcting base stations where radio wave attenu-

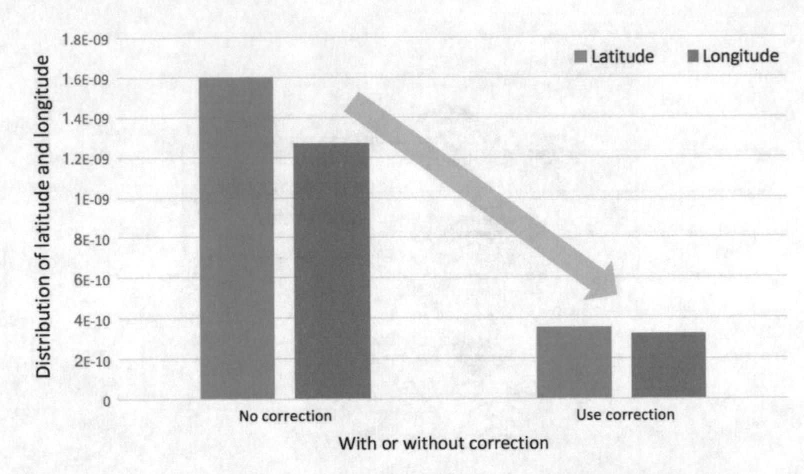

Fig. 5.13 Dispersion comparison of positioning results by presence or absence of correction

ation occurred. Evaluation experiment confirmed that the radio wave attenuation discrimination method is effective irrespective of the number of radio observations. In addition, we showed that when the base station where the radio wave attenuation occurred is known, the outlier of the positioning result is suppressed by the correction by the proposed method, and the variation of the positioning result is reduced. In the future, we will place the particles with the position estimation position as the reference point instead of the grand truth discrimination of the base station where radio wave attenuation occurred, and apply the algorithm. We also examine the number of particles and the range of placement. In addition, in the correction method, instead of correcting with a uniform value, an algorithm is devised which can dynamically correct with an appropriate value from the value calculated by the base station determination algorithm.

References

Beauregard S, Haas H (2006) Pedestrian dead reckoning: a basis for personal positioning. In: Proceedings of the 3rd workshop on positioning, navigation and communication, pp 27–35

Kaji K, Kawaguchi N (2012) Design and implementation of WiFi indoor localization based on Gaussian mixture model and particle filter. In: Proceedings of international conference on indoor positioning and indoor navigation (IPIN). IEEE, pp 1–9

Murata M, Ahmetovic D, Sato D, Takagi H, Kitani K, Asakawa C (2018) Smartphone-based indoor localization for blind navigation across building complexes. In: International conference on pervasive computing and communications (PerCom). IEEE, pp 254–263

Part II
Data Collection and Corpus Construction

Ever since the first studies in activity recognition have shown promising results, a focus has been developed on creating generic and large benchmark datasets with activity data from many study participants and from a variety of sensors. This chapter discusses through five chapters in total, several key lessons that have been learned over the past years in terms of what tools are required to achieve a useful and as generic as possible corpus for activity recognition research.

The first three contributions all examine the annotations of data in particular. Chapter 6 reports from a case study on the detection of drinking gestures from annotations that were made through experience sampling. The dataset in this case study was improved in terms of annotations through two approaches in a post-hoc manner.

The next chapter on the annotation of activity data, Chap. 7, presents a study on the difficulty for nonexperts to design and implement recognition systems that find events in raw sensor data streams. Sensor data and video recordings are combined for example events to create a better interface for labeling examples. The study shows that nonexpert users are able to collect video and sensor data to swiftly and accurately label events using the combined video and sensor data.

In Chap. 8, annotation quality and tools are focused on exploring the challenge of synchronizing several independently recorded sensors. Typical character-separated values (CSV) datasets are known to be slow, cumbersome, and error-prone to exchange. Video recordings, often from multiple angles and time-coded annotations, further complicate curating such data. In this chapter, a possible alternative is presented that uses standardized multimedia formats, in which sensor data are encoded in audio format, and time-coded information, like annotations, as subtitles.

The introduction of specialized tools to deal with activity datasets has lead to work that improves the often-tedious efforts of dataset collection and preparation. In Chap. 9, a toolbox is discussed that provides easy access to datasets by providing them in the same format, and in the same units, measurement range, sampling rates, labels, and body position IDs.

A final chapter in this part touches on the important problem of acquiring more activity data from other sources, which are less challenging in certain scenarios than the usual body-worn inertial sensors. In Chap. 10, a simulator that uses motion capture to simulate accelerometer data on different settings is introduced. The simulated data is then used to estimate the performance of activity recognition models under different scenarios.

Chapter 6
Drinking Gesture Recognition from Poorly Annotated Data: A Case Study

Mathias Ciliberto, Lin Wang, Daniel Roggen and Ruediger Zillmer

Abstract In this chapter we present a case study on drinking gesture recognition from a dataset annotated by Experience Sampling (ES). The dataset contains 8825 "sensor events" and users reported 1808 "drink events" through experience sampling. We first show that the annotations obtained through ES do not reflect accurately true drinking events. We present then how we maximise the value of this dataset through two approaches aiming at improving the quality of the annotations post-hoc. Based on the work presented in Ciliberto et al. (2018), we extend the application of template-matching (Warping Longest Common Subsequence) to multiple sensor channels in order to spot a subset of events which are highly likely to be drinking gestures. We then propose an unsupervised approach which can perform drinking gesture recognition by combining K-Means clustering with WLCSS. Experimental results verify the effectiveness of the proposed method, with an improvement of the F1 score by 16% compared to standard K-Means using Euclidean distance.

6.1 Introduction

Gesture recognition has applications in several fields such as healthcare and sports (Mitra 2007). In order to create a reliable gesture recognition system, it is important

M. Ciliberto (✉) · L. Wang · D. Roggen
Wearable Technologies Laboratory, Sensor Technology Research Centre, University of Sussex, Brighton, UK
e-mail: m.ciliberto@sussex.ac.uk

D. Roggen
e-mail: daniel.roggen@ieee.org

L. Wang
Centre for Intelligent Sensing, Queen Mary University of London, London, UK
e-mail: lin.wang@qmul.ac.uk

R. Zillmer
Unilever R&D Port Sunlight, Birkenhead, UK
e-mail: ruediger.zillmer@gmail.com

N. Kawaguchi et al. (eds.), *Human Activity Sensing*,
Springer Series in Adaptive Environments,
https://doi.org/10.1007/978-3-030-13001-5_6

to have a well-annotated dataset (Bulling 2014). However, creating high-quality datasets may require to rely on lab-like environments, with limited ecological validity (Roggen 2010). Activity recognition research generally strives to employ datasets with unrealistically "perfect" ground truth annotations. In an ecologically valid data collection, however, it is likely that a highly valuable dataset is acquired, but that only poor quality annotations are available.

Experience sampling (ES) is a real-time annotation approach done by users themselves a mobile device (Gjoreski and Roggen 2017). This allows more ecologically valid data collection in everyday life (e.g. no need to video record the experiment). However, ES can lead to the following issues: (i) the synchronisation between the activity performed and the label annotated by the user is generally of poor quality, with the user annotating the activity after the event, or combining multiple activities in a single annotation; (ii) the user may forget to label an event, (iii) the user may annotate an activity with the wrong label.

In this work, we investigate how to make sense of a dataset with high business value comprising drinking gestures, which has been annotated through ES, leading to numerous deficiencies in the annotation quality. The dataset contains drinking gestures annotated by the users with a mobile application. The dataset was collected in an office environment using a 3-axis accelerometer and it is made by 8825 "sensor events", with 1808 "drink events" annotated by users through ES. Using this dataset, we aim to address two main challenges: (i) to understand why the quality of the annotation is low and consequently how would it be possible to improve in future data collection and (ii) to understand whether it is still possible to use such big dataset without relying on the annotations for spotting drinking gestures and how. This work is based on the research presented in Ciliberto et al. (2018). The main contributions are:

- A study of the annotations. We analyse the user annotations, their distribution in time during the data collection, and their relation to the sensor events, in order to understand the causes of the low quality and where the data collection process can be improved.
- A template matching approach, based on Warping Longest Common Subsequence (WLCSS) (Nguyen-Dinh 2012), to extract a subset of drinking gestures, within a certain level of confidence. This subset will allow the dataset to be used for research purposes. As extension of the previous work, in this chapter, we evaluate multiple sensors channel in order to obtain a more precise selection of drinking gestures.
- An unsupervised algorithm (K-Means) adapted to template matching. This algorithm is a new variation of K-Means (Hartigan and Wong 1979) where the WLCSS is used as distance measure. It allows to cluster gestures based on the raw signal of the sensors. At the same time, it clusters gestures taking in account the variation in the way they can be performed, by using WLCSS which has been successfully used for robust gesture detection (Nguyen-Dinh 2012).

6.2 Related Work

The quality of annotations obtained through ES can be poor (Stikic 2009). Annotations issues can include time shift of a label with respect to the activity, as well as wrong or missing labels (Nguyen-Dinh 2014).

Some approaches suggested to improve ES with manual re-annotation (Stikic 2009). This is not feasible economically for a large datasets. Moreover, despite re-annotation the quality may still be insufficient for the training of machine learning algorithms (Stikic 2009). The impact of ES on activity recognition has been studied in Duffy (2018). However, the authors simulated the ES in a controlled environment and they used only the data corresponding to the user annotations.

The problem of poorly labelled data can be tackled during the annotation itself or during the training of the machine learning algorithm. A method useful to reduce the effort of the users while annotating their activity has been proposed in Nguyen-Dinh (2017). The authors suggested a one-time point annotation method that requires the users to only label a single moment per activity rather than specifying the beginning and the end. The method then recognizes automatically the boundary of the activity in the annotated signal. Nevertheless, it requires that the labels are within the execution interval of the activity.

Unlabelled or poorly labelled data are available in big quantities nowadays due to the large diffusion of sensing devices, such as smartphones and wearables devices. For this reason, methods such as semi-supervised learning, active learning and unsupervised learning have been applied in order to extract useful information from sparsely annotated data. A combination of active learning and semi-supervised learning has been studied in Stikic (2008). The authors used a dataset of daily activities collected with two subjects wearing accelerometer sensors and motioned tracked with infrared sensors. This approach however uses a decision window of 30 s long and thus is not suitable for recognizing gestures that occur in a short time. Unsupervised learning has been successfully applied to activity recognition in Kwon (2014) and more recently in Gjoreski and Roggen (2017). In the latter, an activity discovery method based on clustering is proposed to help with ES, although it is designed for periodic movements rather than sporadic gesture. Unsupervised learning has been applied to gestures clustering in Zhang (2011), where a K-Means clustering has been evaluated specifically for hand gestures.

Several studies have tried to address the challenge of activity recognition from poorly annotated data. While most of them used synthetic dataset and focused on periodic or long activity (such as walking, running, etc.), to the best of our knowledge none of them applies to drinking gestures collected in a real-life office environment.

6.3 Dataset

The dataset was collected by providing a set of mugs to 60 users (one mug per user) in an office environment. Each user collected data for a period of 4 days. Each mug was instrumented with a logger comprising a 3-axis accelerometer

(Zillmer 2014). The loggers were placed in a hollow at the bottom of each mug. As the mugs were customly made, the positioning of the loggers was not the same in all mugs. The loggers sample acceleration at 20 Hz, with a timestamp in ms. In order to save power, they start logging acceleration when a movement is detected. After 5 s of inactivity they automatically stop the recording, without record the inactivity period. We use the term *sensor events* to refer to every recording performed by the loggers that lasts at least 4 s (as configured on the loggers for this data collection). Therefore, sensor events can occur for a variety of reasons: moving the mug on the desk, washing it, drinking from the cup, etc.

The data annotation was performed through experience sampling by the users themselves. They labelled each drinking event manually using an Android application installed on their smartphones. Each annotation could be *punctual* or *delayed*. An annotation is considered punctual when it was entered immediately after the drinking event. It is considered delayed, when it refers to an event in the past. The users could specify in the application whether their annotation was punctual or delayed. However they did not have to provide an indication as to how much the delay was. Furthermore, there were no guidance indicating after how much time an annotation should be considered "delayed" rather than "punctual".

The resulting dataset is made by 8825 sensor events, 1808 user annotations, of which 1477 marked as "punctual" and 331 as "delayed". The percentage of annotated gestures with respect to the total amount of sensor events is of 20.5%.

6.4 User Annotation Analysis

We aim to analyse the causes of the poor annotations in order to improve future data collections, as well as helping during the next steps of this study.

Figure 6.1 indicates the main challenge of the annotation protocol, which is how users understood differently how and what to annotate. The data collection protocol did not require participants to annotate drinking solely when using the instrumented mugs: they could annotate drinking as well when using regular mugs. It might happen that users annotated drinking events performed using other cups. The protocol did also not specify what to consider as a "drinking event". Users could interpret it as referring to a single sip, multiple sips, or the act of drinking the entire cup. It is also possible to notice how the annotations are not well aligned with the sensor events.

We also studied the distribution of the labels, per user, over the 4 days of data collection. It could help to understand the users' commitment in annotating their drinking gestures, assuming they were keeping the same drinking habits among all the days. This may be useful in order to spot days for which the annotations can be more reliable. The results are presented in Fig. 6.2. While there is no significant change between day 1 and day 2, with an average increase in the number of annotation of 0.24%, starting from day 3 the engagement decrease by 11% on average among all the users. The plot shows also a great variability in the data: there were users

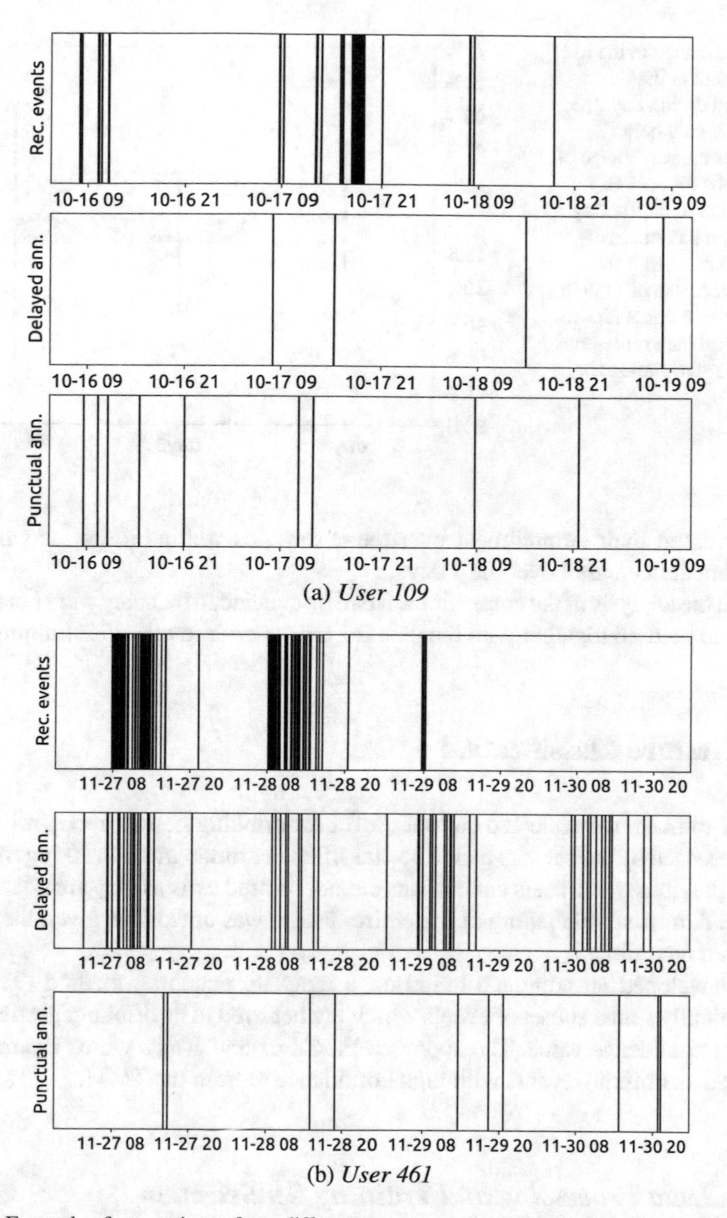

Fig. 6.1 Example of annotations of two different users, over the 4 days period. The start time of the sensor events are displayed in the top plot of each figure, one thin line per event. The delayed and punctual annotations inserted by the users are displayed respectively in the second and the third plot of both figures. The X-axis reports date and time, in the format "MM-DD HH". It is also possible to notice the differences in the way two users annotated the drinking events

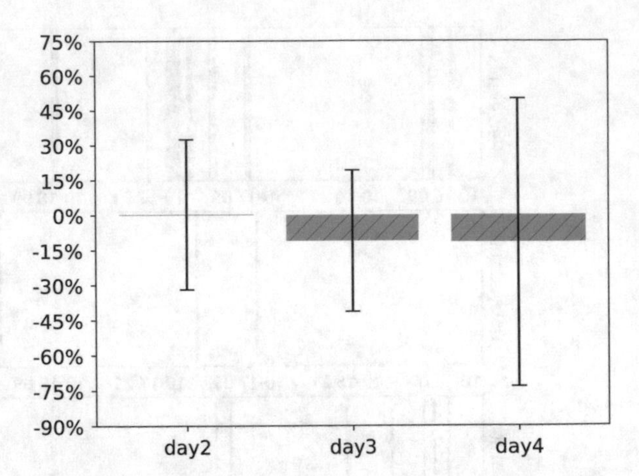

Fig. 6.2 Change in the user commitment in the annotation during day 2, 3, and 4. The grey bars represents the percentage of annotations for each day with respect to day 1. Day 2 displays an increment of 0.24%; Day 3 and 4 an average decrease of 11% in the number of annotations. The vertical bars represents the standard deviation for each day

that increased their commitment over the 4 days, as well as users for which the commitment decreased over the 4 days.

From the analysis of the annotations, it can be concluded that they were not reliable enough to be used together with the data for the supervised classifier training.

6.5 Gesture Classification

In order to make the collected dataset useful for drinking gesture recognition, each event recorded by the sensors had to be classified in drinking/non-drinking. As highlighted previously, the users annotations cannot be used as-is as they are not accurate enough. A manual relabelling of the entire dataset was unfeasible given the lack of any video recordings.

We developed an approach based on a template matching method (TMM) to automatically spot a subset of events which are believed to be drinking gestures with a certain confidence value. The approach then uses few events which are manually identified as drinking events with high confidence to train the TMM.

6.5.1 Data Processing and Training Set Selection

We used a heuristic method to select a few sensor events as the training set. We performed a few drinking gestures using the same instrumented mug in order to chose the best sensor channel for template matching.

The chosen channel (or channels) must be orientation independent, as there was no information about the positioning of the logger in the mug (Fig. 6.3). We discovered that the Z-axis of the accelerometer quite clearly indicates the gesture of lifting the

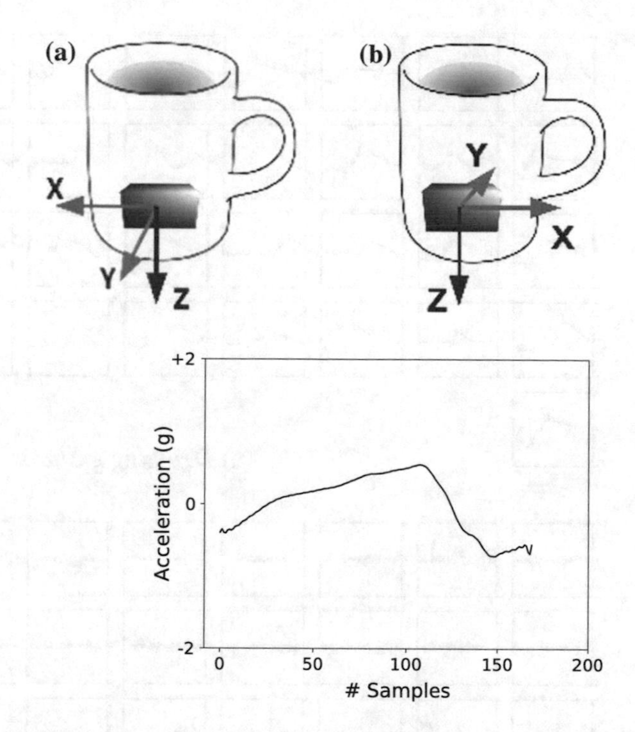

Fig. 6.3 Examples of orientation of the loggers in the custom made mugs. Although X and Y axis can be different in each mug, the Z-axis is always facing downwards

Fig. 6.4 Template of a drinking gesture, performing a single sip, displaying the acceleration on the Z-axis

cup to drink. Also, despite different orientation of the loggers, this axis was always perpendicular to the bottom of the mug. A template of such gesture is displayed in Fig. 6.4. A subset of gestures visually similar to this template was selected manually from the entire set of the available gestures. We selected this subset trying to include some variability in the way the drinking gestures were performed. Another subset of non-drinking gestures was selected too, choosing the templates that were very visually different from the drinking gestures. The training set is displayed in Fig. 6.5. It is formed by 78 events: 37 drinking gestures (Fig. 6.5a) and 41 non-drinking gestures (Fig. 6.5b).

However, lifting the cup does not necessarily mean that a drinking was actually performed. For this reason, in order to better detecting the rotation of the mug due to the drinking, we also used the magnitude of the acceleration on X-Y plane as additional information. This was also due to the lack of the gyroscope on the loggers. The X-Y plane was chosen because it was always parallel to the bottom of the mug (Fig. 6.3). The magnitude was computed as $m_{xy} = \sqrt{x^2 + y^2}$. A template for a drinking gesture represented by the magnitude is shown in Fig. 6.6. A different training set based on the template of the magnitude was chosen. The choice was based on visual similarity to such template. A subset of non-drinking gesture was also chosen for the XY magnitude The training set of drinking and non-drinking gesture for the XY magnitude is displayed in Fig. 6.7. The drinking gestures were selected only when they corresponded to a lifting of the mug (Z-axis).

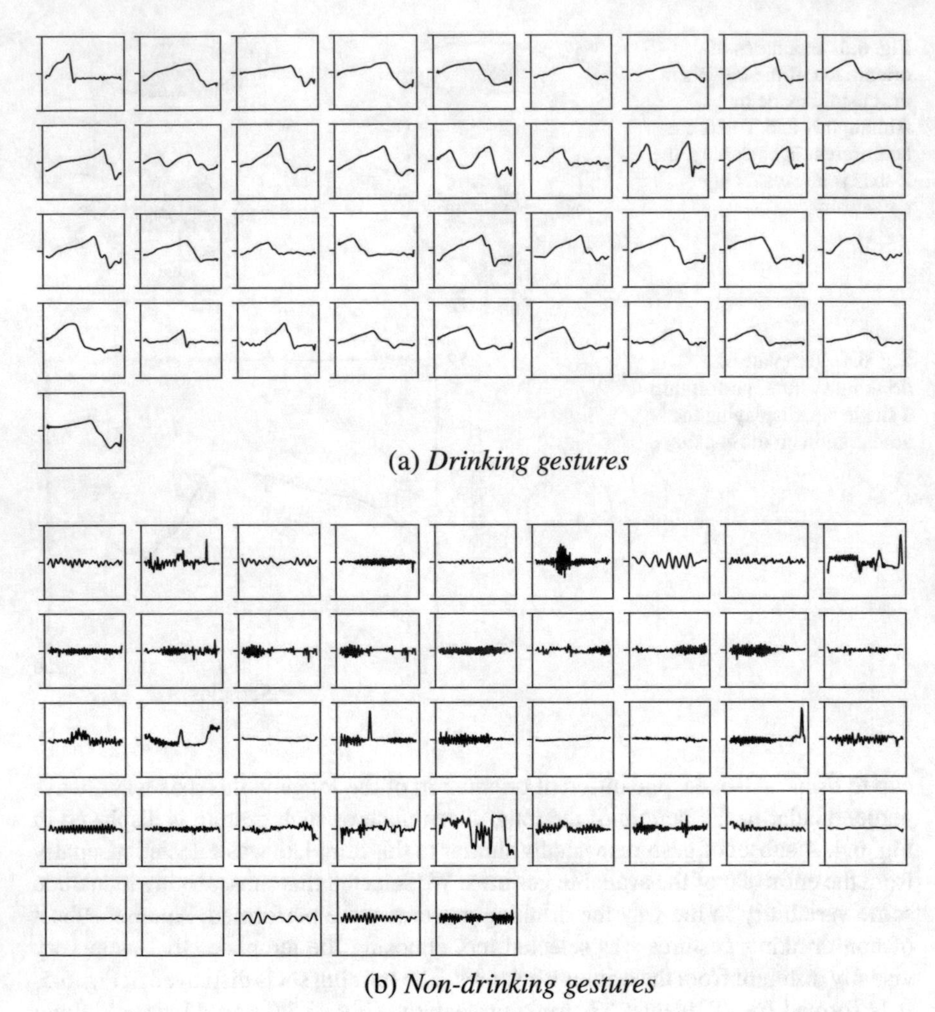

Fig. 6.5 Training set of gestures using the Z-axis of the accelerometer. **a** Displays the templates chosen as drinking gestures. **b** Shows those selected as no-drinking gestures. All the plots show the templates downsampled to the fixed length of 170 samples (X-axis). The Y-axis represents the acceleration within a range of ±2g

All the instances in the dataset were filtered using a Butterworth low pass filter with cut off frequency set to 10 Hz. They were also resampled to a fixed number of samples. The number of samples was selected as the average length of a drinking gesture, which is 170. This step was performed in order to reduce the impact of non-drinking events that can last longer time than drinking gestures (e.g. washing the cup, moving the cup around the office, etc.).

In order to evaluate which is the best channel for drinking gesture recognition, we decided to match the templates for Z-axis signal and XY magnitude separately.

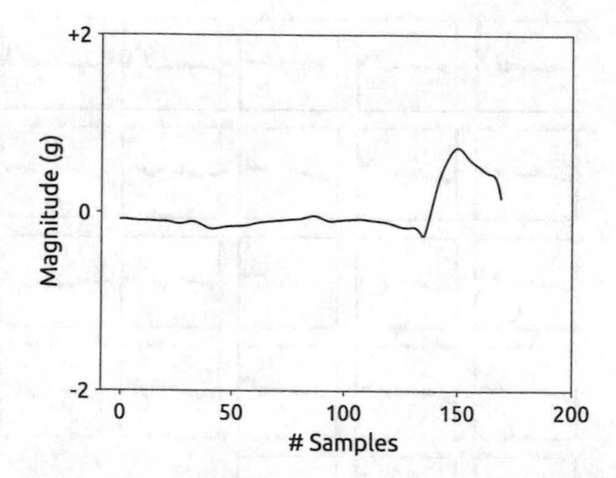

Fig. 6.6 Template of a drinking gesture, performing a single sip, displaying the magnitude of the acceleration on the X-Y plane

6.5.2 *Template Matching Using WLCSS*

The Warping Longest Common Subsequence (WLCSS) (Nguyen-Dinh 2012) is an algorithm developed for template matching in real-time applications. Using dynamic programming, the algorithm can compute a matching score between a template and a stream, updating it at every new sample of the stream. It can be used for gesture recognition as it can handle gestures performed with variation in their speed of execution. This is achieved by three parameters: reward (R), penalty (P) and acceptance distance (ϵ). The algorithm is shown in (6.1).

$$M(i,j) = \begin{cases} 0 & \text{if } \; i \leq 0 \;\; or \;\; j \leq 0 \\ M(j-1, i-1) + R & \text{if } \; |S(i) - T(j)| \leq \epsilon \\ max \begin{cases} M(j-1, i-1) - P \cdot |S(i) - T(j)| \\ M(j-1, i) - P \cdot |S(i) - T(j)| \\ M(j, i-1) - P \cdot |S(i) - T(j)| \end{cases} & \text{if } \; |S(i) - T(j)| > \epsilon \end{cases} \tag{6.1}$$

The matching score M(i, j) is computed as function of the previous scores, by adding a reward (R) when the distance between the i-th stream sample ($S(i)$) and the j-th template sample ($T(j)$) is below an acceptance distance (ϵ), or by subtracting a penalty (P) proportional to the distance, when this is above ϵ. In addition to R, P, and ϵ, WLCSS needs a threshold T. As WLCSS computes a matching score (M) between an instance in the dataset and a template, T is used to define whether an instance matches with the template (Mv $\geq$ T) or not (M $<$ T). The value of the threshold is related to the specificity and the sensitivity of the matching algorithm: a high threshold means high specificity, while a low threshold means high sensitivity. The values of R, P, ϵ and T must be found during the training phase. We optimized this based on an evolutionary optimization technique.

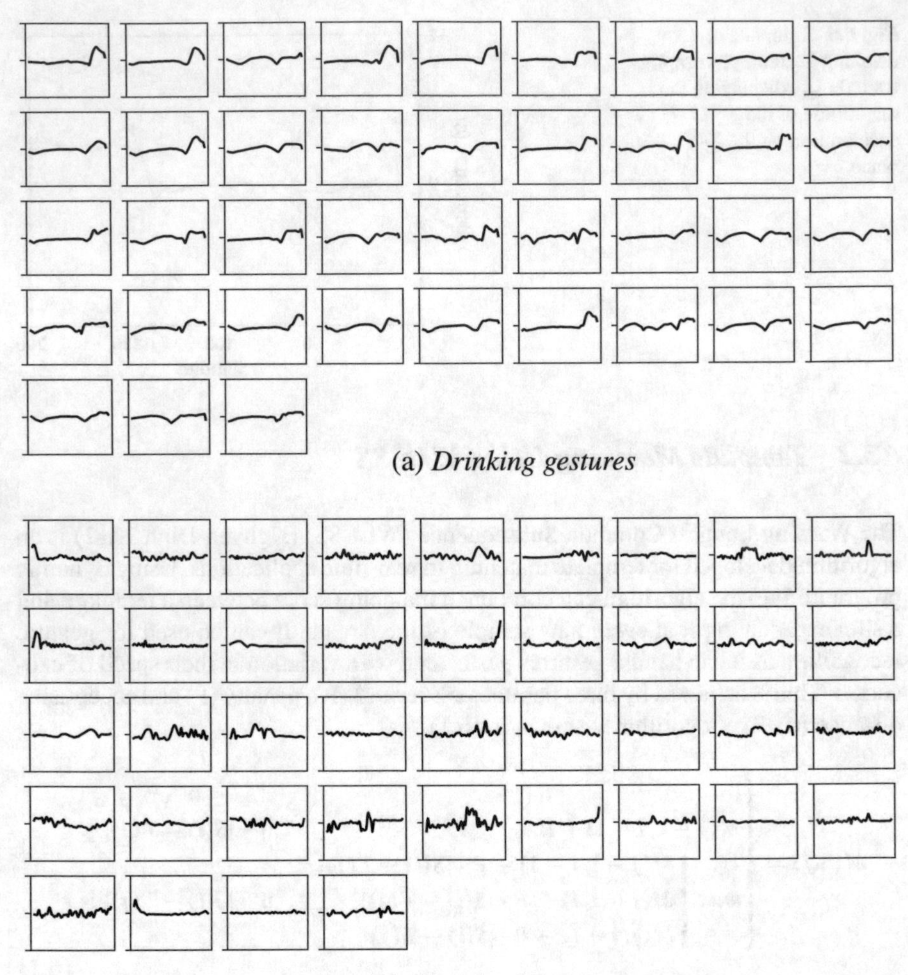

(a) *Drinking gestures*

(b) *Non-drinking gestures*

Fig. 6.7 Training set of gesture using the magnitude of X and Y axis of the accelerometer. **a** Displays the templates chosen as drinking gestures. **b** Shows those selected as no-drinking gestures. All the plots show the templates downsampled to the fixed length of 170 samples (X-axis). The Y-axis represents the acceleration within a range of $\pm 2g$

6.5.3 WLCSS Optimization Using Evolutionary Algorithm

We optimise the values of R, P, e, T to maximise the ability of WLCSS to distinguish drink from non-drink using an evolutionary algorithm (EA). We used the EA in order to optimize the values of the parameters starting from a randomly generated population. Each individual of the population is an array containing the 4 parameters. The EA evolves this population through the usual selection, mutation and crossover operators (Michalewicz 1996). Here, the F1 score is used for the selection. The optimization process stops after a predefined number of iterations, in this case.

6.5.4 Confidence Computation

Given the unreliability of the labels in the dataset, it is not possible to evaluate precisely the correctness of a match for this particular dataset. For this reason, we opted to provide a confidence level for each gesture as output of our method rather than a simple match/no-match. The confidence level was assigned using Four Parameter Logistic Regression:

$$y = d + \frac{a - d}{1 + (\frac{M}{T})^b}$$

where y is the confidence level, M is the matching cost, and T is the threshold. The function, displayed in Fig. 6.8, provides a confidence value in the range [0:1]. The range is defined by the parameters $a = 0$ and $d = 1$. The parameter b, which define the slope of the function curve, was set manually to 5. Using a fixed interval for the confidence makes its value unrelated from the absolute value of T, which can vary according with the parameters R, P and ϵ.

Finally, we computed three confidence values: (i) a value for the Z-axis signal (c_z) used to detect the lifting of the mug, (ii) a value for the matching of the XY magnitude template (c_{xy}) useful for the detection of the mug's rotation and (iii) a combined value that takes into account the previous twos (c_{comb}). This latter value was empirically defined as:

$$c_{comb} = 0.7 * c_z + 0.3 * c_{xy}$$

The weights for c_z and c_{xy} were chosen experimentally based on the assumption that a drinking gesture, intended as rotation of the mug, is performed only after the lifting the mug.

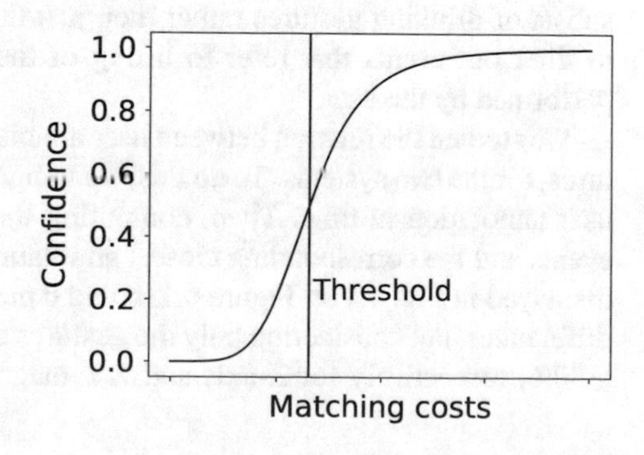

Fig. 6.8 4 Parameter Logistic Regression function used to assign a confidence with the respect to the threshold

6.5.5 Evaluation

We used the two different subset of instances (Z-axis signal and XY magnitude) to train two separate instances of the system using the EA. As the EA is a stochastic process, we repeated the training 10 times, and we picked the best values for each of the two systems. The values are R = 68, P = 0, $\epsilon = 28$ and T = 3364 for the Z-axis and R = 23, P = 11, $\epsilon = 11$ and T = 726 for XY magnitude. With these parameters, we run the algorithm on the entire dataset, by using the templates displayed in Figs. 6.4 and 6.6, which were selected manually from the training sets as templates.

Figure 6.9 displays a comparison of the sensor events and the annotations for a single user, with the corresponding matching scores, for the Z-axis (Fig. 6.9a) and XY Magnitude (Fig. 6.9b). The matching of the template for XY magnitude is less restricting than the matching on the Z-axis. This is possibly due to the fact the rotation of the mug can happen even in case of gesture different from just the drinking: e.g. while washing the mug, moving it around, etc.

The percentages of total detected gestures for each scenario, compared to the total number of events are displayed in Table 6.1, for different confidence values. The low percentages are due to the nature of the sensors, which were collecting all sort of movements such as moving the mug on the desk, washing it or even accidental movements. It is important to note that the number of detected events is also lower than the number of user annotations (1808). This may be a result of the data collection protocol which did not specify to annotated only the drinking movements performed with the instrumented mug. Finally, for low confidence values ($\geq$25% and $\geq$50%) the matching of the Z-axis signal and the XY magnitude are quite different in term of percentage. For higher values of the confidence ($\geq$75%), the performance of the two systems tend to be the same, with similar percentages of detected events. Moreover, in order to better compare the performance of the two systems, we created the cross-correlogram presented in Fig. 6.10. It displays the distribution of the differences between the confidence c_z and $c_x y$. While the two systems generally agree, as shown by the peak on 0, there is a negative skewness. This means that system for XY magnitude is less precise in detecting the drinking events than detecting the lifting of the mug using the Z-axis.

From the analysis, it is clear how the Z-axis signal is better for detecting a first subset of drinking gestures rather then just the XY magnitude. The latter is useful to filter out events that refer to lifting of the cup but without an actual drinking performed by the user.

We studied the relation between user annotations, sensor events and detected gestures, for the two systems. To do this, we assigned to every recorded event the closest user annotation in time. Then, computing the time difference between the sensor events and the corresponding closest annotations, we created the cross-correlogram displayed in Fig. 6.11a. Figure 6.11b and c present the distribution of the same time differences, but considering only the gestures detected by WLCSS with a confidence $\geq$50%, respectively for Z-axis and XY magnitude. The plot for Z-axis presents a

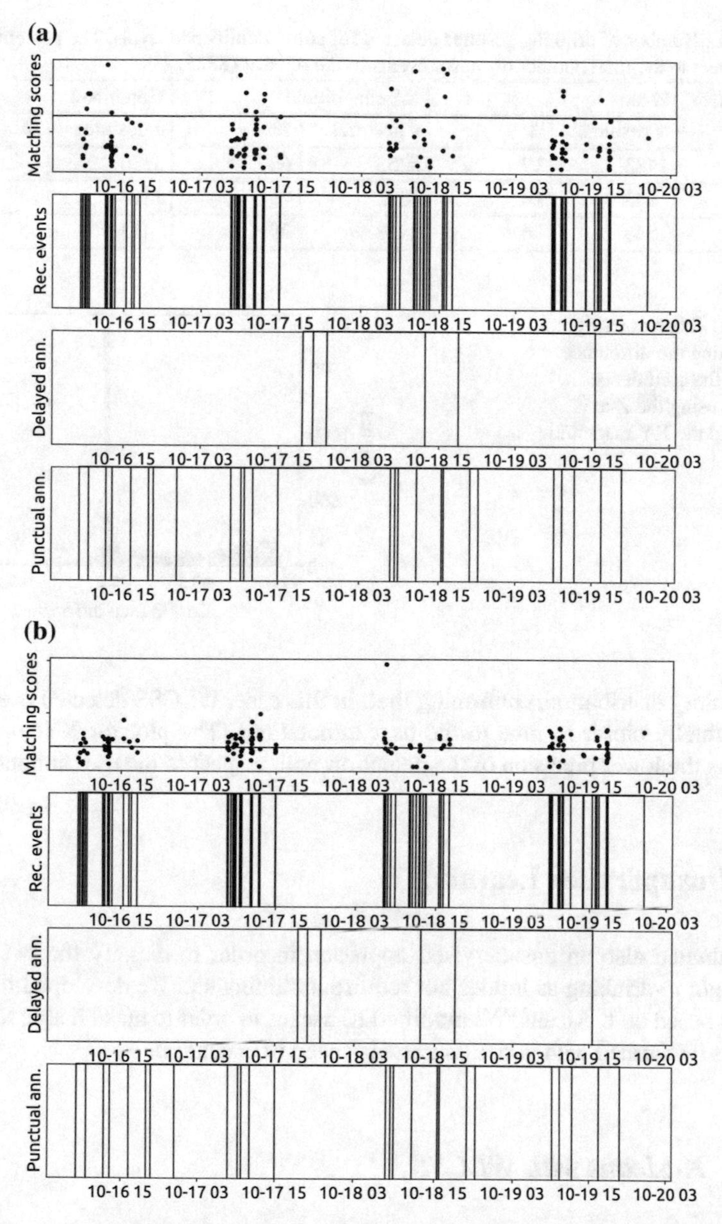

Fig. 6.9 Comparison of WLCSS matching scores with recorded data for a single user, for Z-axis (**a**) and XY magnitude (**b**). For each of the two template matching, from the top: the WLCSS matching scores and the threshold T, as horizontal line (first plot), the start time of the sensor events (second plot), and the user annotations delayed and punctual (respectively third and fourth plot). The data are for 4 days period, with the X-axis reporting date and time in the format "MM-DD HH"

Table 6.1 Number of drinking gestures detected for some confidence levels. The percentages are with respect to the total number of sensor events in the dataset (8825)

Confidence	Z-axis		XY-magnitude		Combined	
	# gestures	%	# gestures	%	# gestures	%
≥25%	1481	17	5253	60	3930	44
≥50%	942	10	4365	49	1113	12
≥75%	543	6	3453	39	483	5

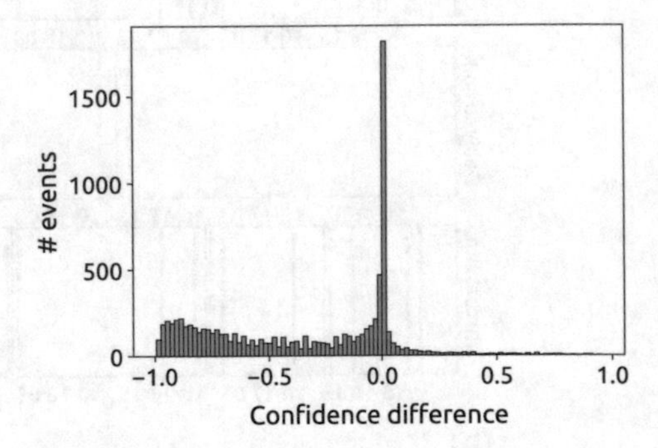

Fig. 6.10 Cross-correlogram representing the difference between the confidence obtained using the Z-axis signal and the XY magnitude

more pointy distribution, confirming that, in this case, WLCSS detected events that were actually closer in time to the user annotations. The plot for XY magnitude confirms the lower precision of the detection with respect to the user annotation.

6.6 Unsupervised Learning

We evaluated also an unsupervised approach in order to classify the gestures in drinking/non-drinking as it does not require a training set. We developed a custom method based on K-Means. We modified K-Means in order to make it able to cluster gestures performed with variation in their speed of execution.

6.6.1 K-Means with WLCSS

K-Means is a clustering technique that aims to partition n observation in k clusters. Each observation belongs to the cluster with the closest mean. It can be used for unsupervised learning by clustering the input data based on a distance measure. The algorithm is based on two steps, assignment step and update step, which are repeated until a stopping criteria is met (Hartigan and Wong 1979). This criteria can be reaching a maximum number of iterations, the change of the clusters in the update step in below a thresholds, etc. We implemented a modified version of the K-Means,

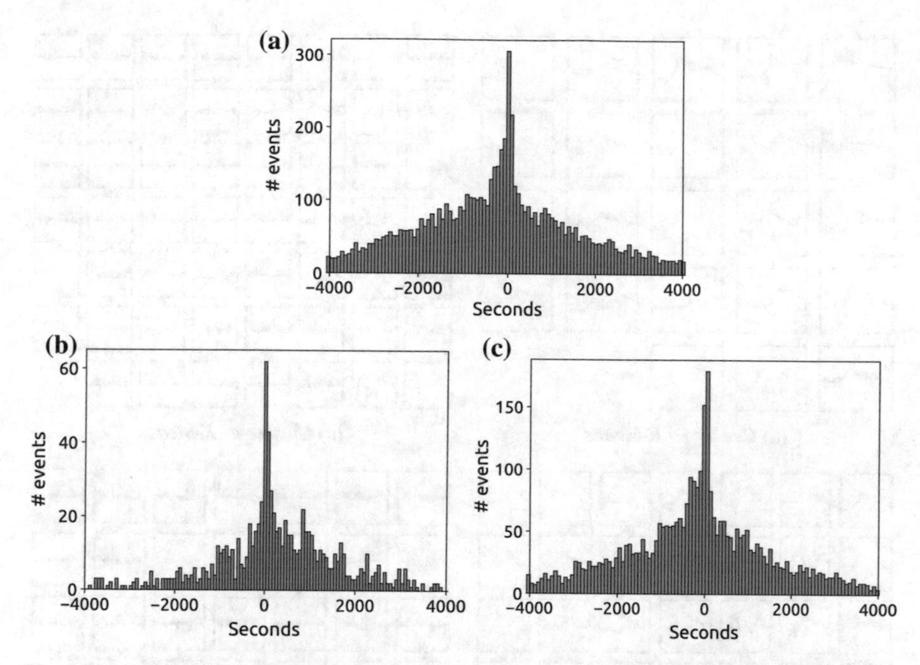

Fig. 6.11 Cross-correlogram representing the distribution of the delays (in seconds) between the user annotations and all the events recorded by the loggers (**a**). Cross-correlogram representing the distribution of the delays (in seconds) between the user annotations and the events detected with a confidence ≥50%, using the signal from the Z-axis only (**b**). Cross-correlogram representing the distribution of the delays (in seconds) between the user annotations and the events detected with a confidence ≥50%, using the XY magnitude (**c**)

where WLCSS was used a distance measure in place of the Euclidean distance. The assignment in our implementation is modificd as following:

$$\arg\max_{c_i} WLCSS(x, c_i)$$

where x is a sensor events, c_i is the centroid for the i-th cluster. The function *argmin* is replaced by *argmax* as WLCSS compute a matching score rather than a distance. The update step is unmodified.

6.6.2 Evaluation

We compared our version of K-Means (named K-MeansWLCSS) against the standard version that uses the Euclidean distance in order to assign each instance to the closest cluster. We applied both the implementation on the training set from the previous step, with $k = 2$ as the goal was to distinguish between drinking and non-

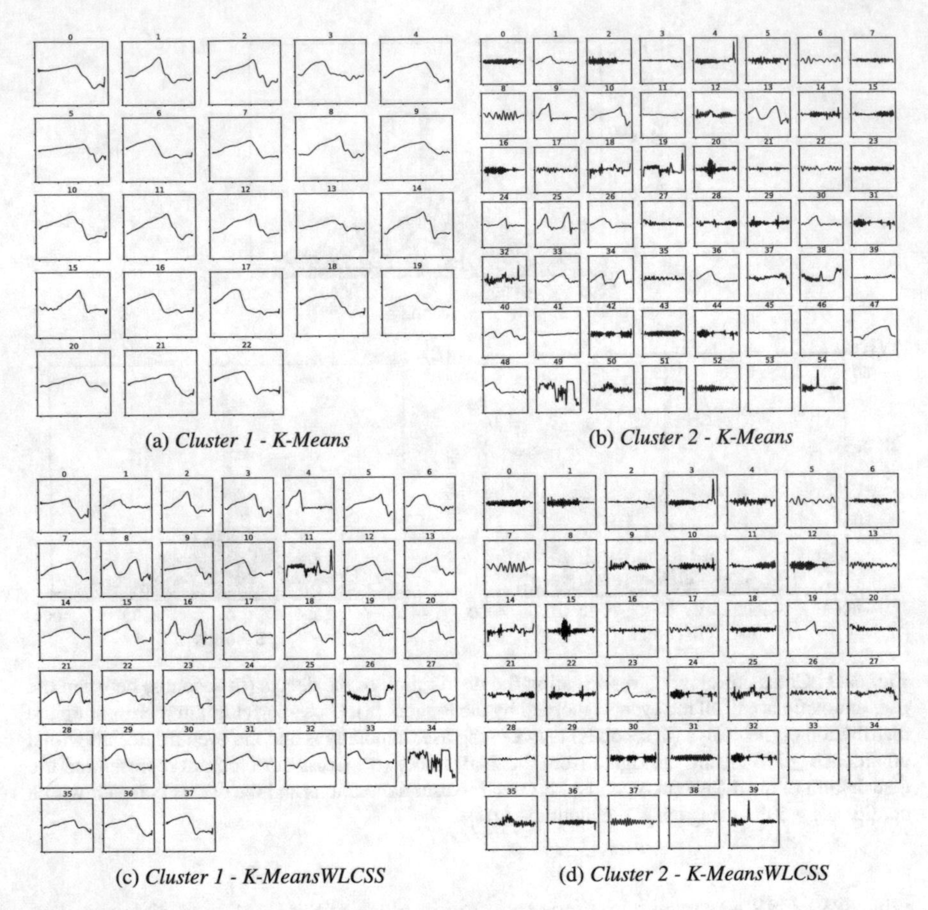

(a) *Cluster 1 - K-Means*

(b) *Cluster 2 - K-Means*

(c) *Cluster 1 - K-MeansWLCSS*

(d) *Cluster 2 - K-MeansWLCSS*

Fig. 6.12 Comparison between clusters obtained with K-Means using Euclidean distance (top left, top right), and using WLCSS (bottom left, bottom right)

drinking gestures. As all the instances were resampled to the same length, they could be used as feature vectors for both the implementations, without dealing with different lengths of the feature vectors (in this case the resampled raw signal). Applying the algorithms on the training set allowed us to compare the clustering results with the labels assigned manually to each gesture, during the data selection. Figure 6.12 presents a visual comparison between the clusters obtained with the two versions of K-Means. For both the implementations, Cluster 1 seems to include mainly the drinking gestures, while Cluster 2 the non-drinking gestures. We used this consideration in order to evaluate the performance of the two algorithms computing precision, recall and F1 score, presented in Table 6.2. They are computed by comparing the clustering of K-Means with the manual labels of the instances in the extracted subset. K-Means_WLCSS increased the F1 score by 16%, being able to detect more

Table 6.2 Precision, Recall and F1 score for K-Means and K-MeansWLCSS computed on the training set. They are computed using the manual labels assigned to each instance in the training set as ground-truth. The majority of the gestures in a cluster is used as classification label for the K-Means implementations.

	K-M (%)	K-M_WLCSS (%)
Precision	100	92.11
Recall	62.16	94.6
F1	77	93

variations in the drinking gestures as it is also visible from Fig. 6.12. It was able to cluster correctly drinking gestures composed by two sips, such as the instances 8, 16, and 21 of Fig. 6.12c.

6.7 Discussion

We discovered that the main issue for this dataset was the data collection protocol which was too relaxed. More precise instructions would increase considerably the quality of the data. Simultaneously, asking the users for more precision in following the protocol should be balanced with shorter sessions of data collection, as we noticed how the user commitment decreases over 4 days of continuous data collection. Lastly, as it has been demonstrated that experience sampling is not reliable, we recommend to increase the effort in the setup of the experiment by including a video recording. It would dramatically increase the quality and re-usability of the dataset, although it would require additional time for the labelling of the data. In order to reduce this effort, the video recordings can be used to precisely annotate just a small portion of the entire dataset. This well-annotated subset, which can be also collected a-posteriori, or can be used as training set instead of extracting one through heuristic. The well-annotated dataset would also allow to evaluate more precisely the performance of the TMM or any other classifier on the complete dataset, through statistical analysis.

In this study we extracted a subset of events which can be considered as drinking gestures with a certain confidence. This extracted subset can be potentially used to re-train the TMM for a more reliable gesture recognition system. The re-training phase can be performed using gestures with different levels of confidence: an higher level of confidence would increase the specificity of the found gestures. Decreasing this value would increase the sensitivity, potentially including more variations of the drinking gestures.

Finally, we aimed to evaluate how an unsupervised learning technique can be used in order to extract drinking gestures from a poorly labelled dataset. We implemented a modified version of K-Means which uses WLCSS as distance measure for the assignment step. The results are promising: with 2 clusters it managed to differentiate between drinking and non-drinking gesture with an 93% F1-score, although on a limited number of sensor events. A more extensive evaluation can be performed on the entire dataset, although without a reliable ground-truth, a validation of the results, in this case, could be difficult.

6.8 Conclusion

In this study, we investigate how to extract knowledge from a poorly labelled dataset of drinking gestures. We analysed the user annotations in order to get qualitative information on how to improve the data collection. We exposed how the loose protocol created most of the problems and we highlighted the need of providing more precise instructions to the users. Then, by selecting instances manually and using a template matching algorithm, we demonstrated that it is possible to extract a subset of instances which are actually drinking gestures within a certain level of confidence. We proved that an unsupervised approach based on K-Means and WLCSS can improve the clustering of gesture over the standard K-Means implementation. Our method outperformed the baseline method by including a wider variety of drinking gestures and increasing F1-score by 16%.

References

Bulling A et al (2014) A tutorial on human activity recognition using body-worn inertial sensors. ACM Comput Surv (3):1–33

Ciliberto M, Wang L, Roggen D, Zillmer R (2018) A case study for human gesture recognition from poorly annotated data. In: Proceedings of the 2018 ACM international joint conference and 2018 international symposium on pervasive and ubiquitous computing and wearable computers. ACM, pp 1434–1443

Duffy W et al (2018) Addressing the problem of activity recognition with experience sampling and weak learning. In: Proceedings of SAI intelligent systems conference, pp 1–6

Gjoreski H, Roggen D (2017) Unsupervised online activity discovery using temporal behaviour assumption. In: Proceedings of the ACM international symposium on wearable computers, pp 42–49

Hartigan JA, Wong MA (1979) Algorithm AS 136: A k-means clustering algorithm. J R Stat Society Ser C (Appl Stat) (1):100–108

Kwon Y et al (2014) Unsupervised learning for human activity recognition using smartphone sensors. Expert Syst Appl (14):6067–6074

Michalewicz Z (1996) Evolution strategies and other methods. Genetic algorithms + data structures = evolution programs. Springer, Berlin, Heidelberg, pp 159–177

Mitra S et al (2007) Gesture recognition: A survey. IEEE Trans Syst Man Cybern Part C: Appl Rev (3):311–324

Nguyen-Dinh LV et al (2012) Improving online gesture recognition with template matching methods in accelerometer data. In: International conference on intelligent systems design and applications, pp 831–836

Nguyen-Dinh LV et al (2014) Robust online gesture recognition with crowdsourced annotations. J Mach Learn Res 3187–3220

Nguyen-Dinh LV et al (2017) Supporting One-Time Point Annotations for Gesture Recognition. IEEE Trans Pattern Anal Mach Intell (11):2270. http://ieeexplore.ieee.org/document/7778186/
Roggen D et al (2010) Collecting complex activity datasets in highly rich networked sensor environments. In: Proceedings of international conference on networked sensing systems, pp 233–240
Stikic M et al (2008) Exploring semi-supervised and active learning for activity recognition. In: IEEE international symposium on wearable computers, pp 81–88
Stikic M et al (2009) Activity Recognition from Sparsely Labeled Data Using Multi-Instance Learning. In: International symposium on location- and context-awareness. IEEE, pp 156–173
Zhang X et al (2011) A framework for hand gesture recognition based on accelerometer and EMG sensors. IEEE Trans Syst Man Cybern Part A: Syst Humans (6):1064–1076
Zillmer R et al (2014) A robust device for large-scale monitoring of bar soap usage in free-living conditions. Pers Ubiquitous Comput (8):2057–2064

Chapter 7
Understanding How Non-experts Collect and Annotate Activity Data

Naomi Johnson, Michael Jones, Kevin Seppi and Lawrence Thatcher

Abstract Inexpensive, low-power sensors and microcontrollers are widely available along with tutorials about how to use them in systems that sense the world around them. Despite this progress, it remains difficult for non-experts to design and implement event recognizers that find events in raw sensor data streams. Such a recognizer might identify specific events, such as gestures, from accelerometer or gyroscope data and be used to build an interactive system. While it is possible to use machine learning to learn event recognizers from labeled examples in sensor data streams, non-experts find it difficult to label events using sensor data alone. We combine sensor data and video recordings of example events to create a better interface for labeling examples. Non-expert users were able to collect video and sensor data and then quickly and accurately label example events using the video and sensor data together. We include 3 example systems based on event recognizers that were trained from examples labeled using this process.

N. Johnson (✉)
University of Virginia, Charlottesville, VA 22903, USA
e-mail: snj3k@virginia.edu

M. Jones · K. Seppi · L. Thatcher
Brigham Young University, Provo, UT 84604, USA
e-mail: jones@cs.byu.edu

K. Seppi
e-mail: k@byu.edu

L. Thatcher
e-mail: lwthatcher@msn.com

N. Kawaguchi et al. (eds.), *Human Activity Sensing*,
Springer Series in Adaptive Environments,
https://doi.org/10.1007/978-3-030-13001-5_7

7.1 Introduction

Small sensors on printed circuit boards (PCBs), such as accelerometers or gyroscopes, are widely available and inexpensive. Many tutorials[1] exist for connecting these small sensors to computing devices like the Arduino[2] or the Raspberry Pi.[3] Using these tutorials, it is not difficult for many people to connect a sensor to a device and to watch the data stream in real time through a console window.

Unfortunately, it remains difficult to write programs that find meaningful events in the data stream. Finding meaningful events requires a person to identify meaningful patterns in the raw data and then to write a program that identifies those patterns. This task is particularly difficult for non-technical people who have little to no experience working with time-varying signals and programming.

Building event detectors for streaming sensor data is important because it can enable the creation of interesting interactive systems. As computing moves from the desktop to the pocket, and now to smaller single purpose devices, event detectors enable devices that respond to events in the world around them. Several physical interactive devices that recognize events in real time from sensor data have recently appeared in the literature (Laput et al. 2015; Zhang and Harrison 2015; Jin et al. 2015).

While many people have ideas for these kinds of physical interactive devices, they may lack the technical skills needed to design and implement the event detector. Reducing the time and effort needed to build an event recognizer will allow faster iterations over ideas for interactive systems.

One way to create an event detector for an interactive system is to use machine learning to train a classifier from example events labeled in data. The gathering and annotation of data for this purpose can be done with tools such as ANVIL, ELAN or ChronoViz; in this chapter we use a "video annotation tool" (VAT) that we created. VAT shows the user video of example events synchronized with the sensor data. In this process, the user records video and sensor data for the example events and then labels the video and sensor data together. The learner takes the labels and the sensor data as input to learn a classifier that operates on the sensor data alone. The video is captured only to help the user during labeling. We also created a data logger on a PCB which we used to collect all sensor data used in this chapter.

Figure 7.1 illustrates the labeling process in the context of a scenario involving an LED turn signal system for cyclists. The system converts bicycle hand signals into LED signals on the cyclist's backpack. First, the designer recruits a friend to make example hand signals while riding a bike. The designer attaches a sensor to the friend's hand and films the friend making example signals. Next, the designer imports the sensor data and the video into our system. The designer defines a set of labels for the events such as "left," "right," and "stop."

[1] See http://learn.adafruit.com/ or https://learn.sparkfun.com/.

[2] http://www.arduino.cc.

[3] http://www.raspberrypi.org.

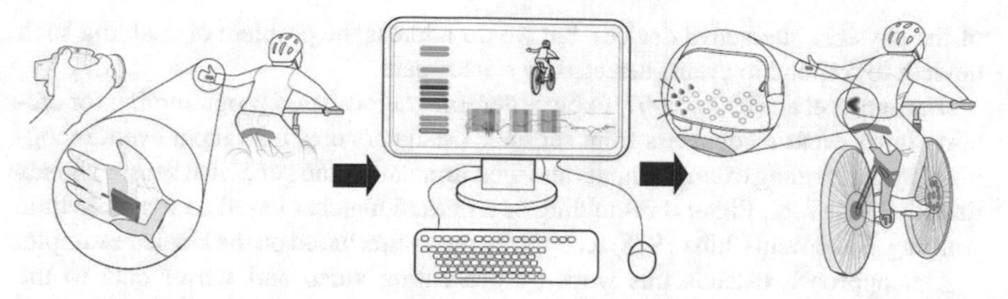

Fig. 7.1 Using VAT to build an event recognizer for bicycle hand signals by labeling example video

Both the video and the sensor data are visible during the labeling process which is shown in the center of Fig. 7.1. The designer can use the video and sensor data to quickly locate and confirm events. The video and data are always kept synchronized so that the current video frame shows what happened for a given sensor value and vice versa. As the video plays, the event stream moves to remain synchronized with the video.

After a while, the designer learns to recognize patterns in the data that correspond to events and can click directly on the next event in the data stream to skip through the video. The video confirms that what the designer thought was an event based on the data is actually an event.

After labeling, the designer passes the label set and the data stream to a machine learning algorithm which learns an event recognizer. The recognizer can then be incorporated into a working prototype of the system as shown on the right in Fig. 7.1.

These steps outlined above were discussed in a previous paper with the same title (Jones et al. 2018). This chapter extends our prior work in three ways: first, by providing more details about how novices can build an event recognizer; second, by providing an in-depth discussion of how to adapt the GMM-HMM approach for event recognition; and third, expanded results that include a study where we have evaluated the use of video and data together for non-experts to label events in the context of 4 studies involving 3 different applications with 74 participants.

7.2 Related Work

7.2.1 Interactive Physical Devices

Building event recognizers fills an important gap in the construction of physical interactive devices. Prior work addresses the physical shape of the object (Savage et al. 2015a; Hudson and Mankoff 2006), and adding functional physical widgets to the object (Savage et al. 2015b; Harrison and Hudson 2009), data collection and event recognition (Jones et al. 2016, 2018). We do not address the design or construction of the physical interactive devices, but we do address the problem of enabling such devices to respond to events detected by sensor data.

Hartmann et al. (2006, 2007) explore demonstration-based programming for creating input event recognizers from sensors. Designers create an input event recognizer by performing example inputs and then annotating the generated sensor signals directly in the tool. Either thresholding or a pattern matcher based on dynamic time warping (Sakoe and Chiba 1978) recognizes the events based on the labeled example.

Our approach extends this work by presenting video and sensor data to the designer, supporting use outside of the laboratory environment, and by using a different learning algorithm to create an event recognizer. Presenting video and data during labeling decouples example generation from event labeling and allows the system to be used when not connected to a workstation.

7.2.2 *Event Recognizers and Interaction*

Machine learning is becoming a widely used approach to learning classifiers for sensor input. Several recent papers (see Laput et al. 2015; Zhang and Harrison 2015; Jin et al. 2015) develop specific interactive systems by learning to recognize specific events or conditions in sensor data. The work that we present in this chapter is different in that we envision a general purpose system for developing event recognizers, not one-off creation of recognizers for specific applications.

7.2.3 *Hidden Markov Models, ASR and Other Activity Models*

HMMs and other sequence models have been applied to activity recognition (Bulling et al. 2014; Patterson et al. 2005) but we employed signal processing for physical computing devices that most closely resembles *automatic speech recognition* (ASR). The translation of acoustic signal into discrete words closely parallels the conversion of motion sensor readings into discrete real-world physical events. One common approach to the ASR problem is to train a *Gaussian-mixture model based hidden Markov model* (GMM-HMM) with the Baum-Welch algorithm and to label recordings using the Viterbi algorithm (Jurafsky and Martin 2009). Fortunately, the strong assumptions undergirding this generative model let it rapidly fit labeled training data (Murphy 2012), making it a good choice for the back end of a fast and focused tool for interactive machine learning. This approach to activity recognition can be seen as a combination of activity recognition based on HMM's (Patterson et al. 2005) and and GMMs (Plötz et al. 2011).

7.3 Building an Event Recognizer with VAT

Building an event recognizer with VAT consists of the following six steps:

1. Define events and split the events into pieces.
2. Attach the data logger to the person or object performing the events.
3. Record video and data of example events.
4. Label the events and pieces of events in the data using the video as a guide.
5. Run the learner on the labels and data.
6. Deploy the learned event recognizer.

We describe the process using a running example in which a designer wants to build a system that recognizes bicycle hand signal events as shown in Fig. 7.1.

The event recognizer discriminates between the different kinds of hand signals so that the correct LED pattern is shown. For example, if the cyclist signals "left turn" by holding their left arm straight out to the left, the system flashes the left turn arrow on the cyclist's back.

7.3.1 Define Event Pieces

Events are decomposed into pieces. A piece of an event is a simple motion that is contained in every instance of an event. The designer might divide the events in the bicycle signal example into two pieces: raise and lower. The "raise" piece represents moving the hand from the bicycle handle bar to the signal position and the "lower" piece represents moving the hand from the signal position back to the handle bar.

The left turn, right turn and stop events all consist of the same pieces (lower and raise) in the same order but the exact motion is different in each case. The learner can distinguish between events represented by the same sequence of pieces as long as the actual motions represented by each event are different.

7.3.2 Attach Data Logger

The data logger is attached to the person or object that will perform the actions so that the data logger can record acceleration and velocity. The data logger should be placed where it will move in ways that are significant for recognizing the event. Figure 7.2 shows the data logger attached to a bicycling glove. Placing the data logger on the back on the hand captures the large motions associated with each signal event. Placing the data logger on the upper arm would not work as well because the upper arm experiences less motion in each of the signals.

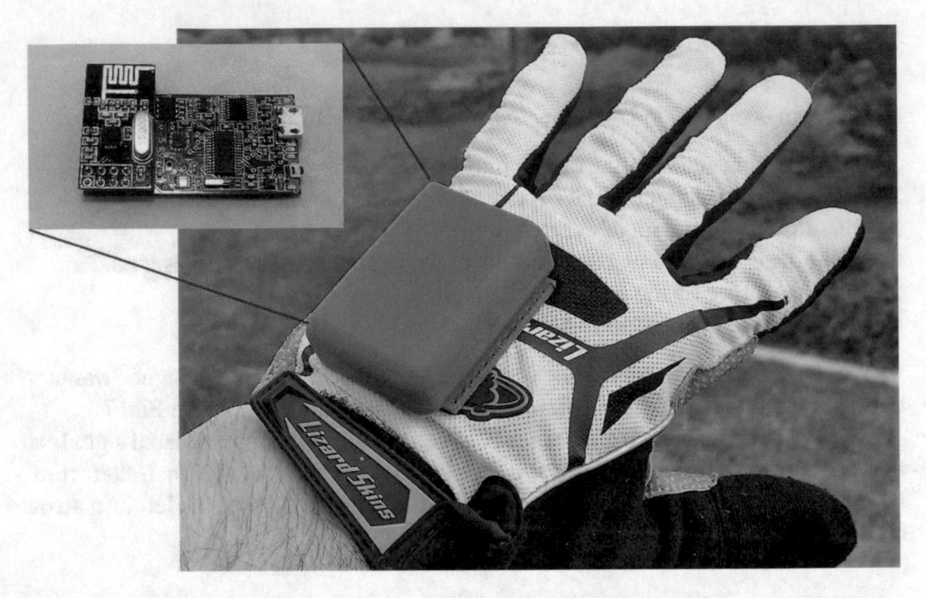

Fig. 7.2 The data logger attached to a bike glove. The data logger is secured in the orange housing

The data logger records acceleration and rotation in 3 axes each.[4] The data are either written to an SD card or broadcast over a wireless link to a receiver. The sampling rates for acceleration and rotational velocity can be adjusted independently using a configuration file. In most cases, we sample acceleration and rotation each at 25.6 samples per second. The data logger will support sampling rates upto 102.4 samples per second for the accelerometer when writing to the local SD card. Other sensors could be used with VAT as long as the data sensed results in some visible or audible change that can be captured on video.

7.3.3 Record and Synchronize Video and Data

The user records video and data of the events. We have developed a process to simplify synchronization of video and data upto the video camera framerate.

The synchronize the video and data, the user captures a red flash generated by the data logger ten seconds after the data logger is turned on. The data logger does not need to be visible after the first red flash is captured. The data logger logs the time of the first red flash (in milliseconds elapsed since the processor was turned on). At the beginning of labeling in VAT, the user locates and marks the first video frame that contains the first red flash. VAT can then synchronize the video and data for as long as both continue recording. A red line over the sensor data display indicates the

[4]Using an Invensense MPU-9250.

data that corresponds to the currently displayed video frame. If either is stopped or turned off, synchronization must be repeated.

The accuracy of this approach is limited by the video camera frame capture rate. For the examples used in this chapter, we captured video at 30 frames per second which means that synchronization of the video and data is off by at most 0.033 of a second. Cameras with higher frame rates can be used to synchronize the data with greater accuracy (up to the temporal resolution of the data logger sampling rate). But for human generated events considered in this chapter, 0.033 of a second proved sufficient.

7.3.4 Label Events in VAT

Figure 7.3 shows the use of VAT for labeling. After marking the frame containing the first red flash, the user is presented with the video, the data stream, and a list of buttons. The buttons are used to label events and pieces. Video playback can be controlled using the space bar to toggle play and pause and the arrow keys to advance a frame forward or backward.

The data stream display is synchronized with video playback so that the red vertical line in the data stream display is always over the data recorded during the current video frame. The data stream is also interactive. Clicking in the data stream advances video playback to the corresponding location in the video; scrolling the mouse wheel zooms in and out of the data stream time scale. The numbers beneath

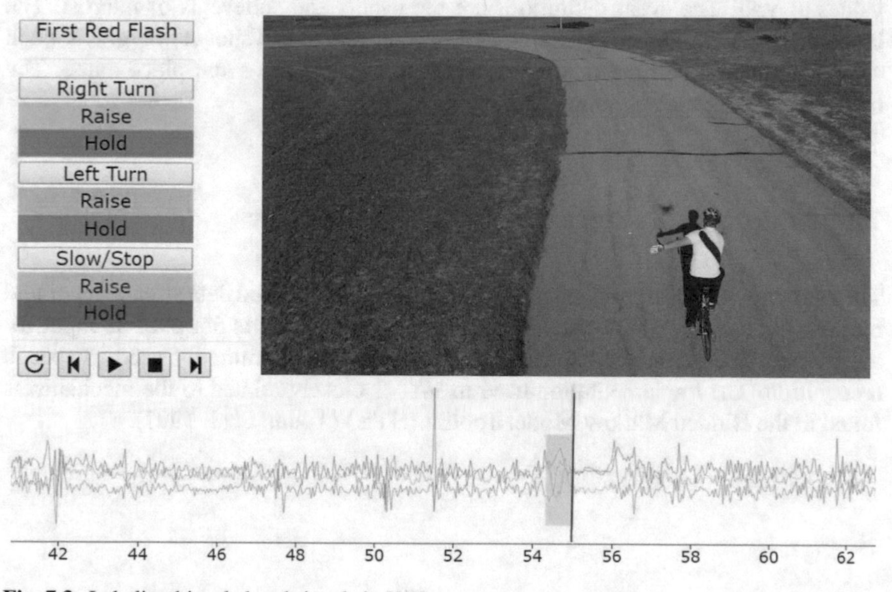

Fig. 7.3 Labeling bicycle hand signals in VAT

the data stream display represent the number of seconds since the data logger was turned on.

In Fig. 7.3, the user is in the process of labeling the “raise” piece of a left turn event. The orange highlight in the data display window marks the region of data currently labeled “raise.” The highlight color corresponds to the color of the label shown under the “Left Turn” button.

In the frame of video shown in Fig. 7.3, the rider has extended his arm to the left and will hold it stationary for about one second (from second 55 through second 56) and then put his hand back on the bicycle handlebars. These actions can be seen in the data stream. The acceleration to the left of the red playback line, which is highlighted in orange, corresponds to the rider lifting and straightening his arm. The region to the the right of the red playback line has small acceleration values due to rider holding his arm stationary and extended. After the rider places his hand back on the handlebars, the data logger again begins to record high frequency acceleration due to vibrations transmitted from the bike to the rider’s hand.

After assigning labels while viewing both the video and the data, most users begin to recognize patterns in the data associated with events. At that point, the user can quickly click through the data stream to identify likely events and use the video to confirm each event and carefully locate the boundaries between event pieces.

7.3.5 Train Recognizer

The user trains an event recognizer by passing the event definitions, labels and data values to VAT. The event definitions are the events and subevents of interest. The labels are a list of 4-tuples generated as described above and shown in Fig. 7.3. Each 4-tuple contains two timestamps, the event name, and the event piece name. The timestamps mark the beginning and end of the event piece.

7.4 Learning an Interactive Event Recognizer

The learner in VAT learns an event recognizer from the labeled data stream generated by the user; the video was only used to guide the user and is not used as input for the learner. The learner is based on the GMM-HMMs commonly used in speech recognition. The implementation used in VAT is closely related to the mechanisms found in the Hidden Markov Model Toolkit (HTK) (Young et al. 1997).

7.4.1 GMM-HMM Approach for Speech Recognition

In ASR, the recognizer identifies words and word boundaries as they occur in an audio signal. This matches the problem of event recognition using the data logger because events are sensed as a time-varying signal. ASR algorithms are also designed to be robust to variations in the speed and way words are spoken by different people. Similar robustness is needed to recognize events performed by different people.

Speech is often represented using a hidden Markov model, with the hidden nodes representing parts of phonemes and the observed nodes representing segments of the discretized acoustic signal. Typically each phoneme is represented with a sequence of three of hidden nodes with each node representing the beginning, middle, and end of the phoneme (Jurafsky and Martin 2009). The observed nodes can be implemented using the (logarithm of the) of the product of Gaussian mixture models (GMMs). Using a GMM-HMM is convenient because the hidden node transition model can be learned by Baum-Welch and the GMMs can be learned by k-means clustering (Young et al. 1997, p. 17); the resultant model can then be used to decode spoken words by the Viterbi algorithm (Young et al. 1997, p. 9).

7.4.2 Adapting the GMM-HMM Approach for Event Recognition

We adapt the structure and algorithms used in ASR to the problem of recognizing real-world events from sensor signals in VAT. To adapt these ASR algorithms to event recognition, we divide the event to be recognized into pieces in the same manner that spoken words are split into phonemes. A phoneme is a phonetic piece of a word. For example, the word cool has three phonemes with each phoneme corresponding to each spoken sound in the word. For events, each piece of the event is a specific motion that is performed as part of an event.

Figure 7.4 shows a scoop event (as might be found in scooping an ingredient for cooking) split into four pieces: a "load" in which the wrist rotates the spoon to load

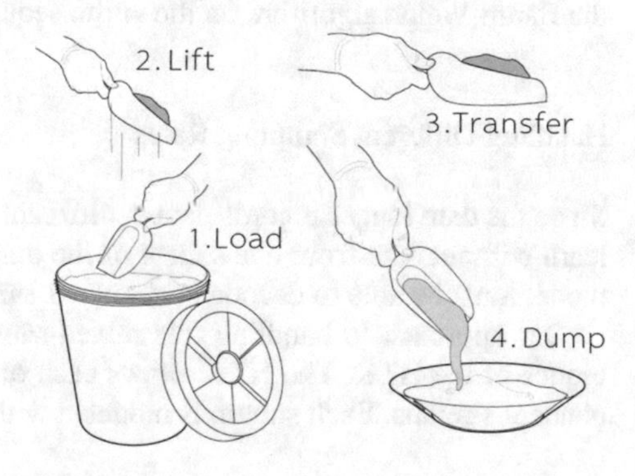

Fig. 7.4 Four pieces of a scoop event: loading, lifting, transferring and dumping

the spoon, the "lift" in which the spoon is raised out of the container, followed by a "transfer" in which the hand moves the spoon, and ending with a "dump" in which the wrist rotates again to the dump the contents from the spoon. Each scoop event contains these four pieces in that order though different people may do each piece differently. We also include a "nothing-happening" event which is applied after the user is done labeling events to mark time periods in which the user did not label any event.

Just as phonemes can occur in different words, motions can also appear in different contexts and with different HMM transition probabilities. In the scooping example, the "transfer" motion might also appear as a piece of the "wash spoon event."

We adapt GMM-HMMs to the problem of event recognition by modifying the process for learning transition probabilities, handling data streams with different sampling rates, and adding an event length threshold. Each adaptation is discussed below.

Learning Transition Probabilities

ASR researchers often represent each phoneme with three hidden nodes; similarly, we represent each piece of an event with three hidden nodes. For the scooping example, this means that each of the four pieces of the scoop event (load, lift, transfer, and dump) are each represented by three hidden nodes; thus, including the hidden node for the "nothing-happening" event, we have a total of $(4 \times 3) + 1 = 13$ hidden nodes. In general, the HMM for an event has $(n \times 3) + 1$ hidden nodes, where n is the number of event pieces.

Recall that a transition probability in an HMM is the probability of moving from one hidden node to another. Because the user does not observe the transitions between hidden nodes, the transition probabilities between the three hidden nodes for a single event piece are learned by uniform segmentation and then Viterbi segmentation as in the HTK (Young et al. 1997, p. 142). The user does observe and mark transitions between event pieces, so these transitions are learned directly from the labels. Combining the transition probabilities within event pieces with the transition probabilities between event pieces gives a full hidden-node transition model without using the Baum-Welch algorithm on the entire sequence.

Handling Different Sampling Rates

Since the data logger records data at different rates, our learning algorithm needs to learn without data from one sensor or the other at some time steps, and the learned model must be able to decode such mixed-sampling-rate input data.

Our approach to handling this mixed-sampling-rate data builds upon the techniques of the HTK. The HTK allows each observation vector to be split into independent streams. Each stream is modeled with a GMM and the weighted product of these stream GMMs is the observation probability density for a hidden node (Young et al. 1997, p. 6).

Our approach also departs from the HTK approach in the selection of the number of mixture components. Like some ASR researchers (Chen and Gopalakrishnan 1998), we select the number of mixture components in the GMM by optimizing the Bayesian information criterion (BIC), rather than using a fixed user-specified number of mixture components. We use this BIC-optimization approach because our users are likely unfamiliar with GMMs and the ideal number of mixture components varies, with some dependency on the number of training examples. To limit the training time, we cap the number of mixture components at three.

Event Length Threshold

Given the transition probability model and the observation probability models both learned as described above, the Viterbi algorithm can be used to classify events in the application. However, the Viterbi algorithm does not consider the duration or length of the event. This means that the algorithm occasionally incorrectly classifies a very short sequence of data as an event. For example, in the scoop data the algorithm might incorrectly label a sequence of noise in the data with duration 0.3 s as a scoop event. Human generated scoop events do not happen at that time scale.

We set the event length threshold to the duration of the shortest labeled event in the training data minus two standard deviations. We also set the minimum event piece length threshold to 0.05 s because motion at that time scale can not be reliably detected in data captured at less than 20 samples per second. Video captured by a high speed camera, and data captured at a higher sampling rate, may support event pieces with shorter durations.

7.5 Results

We have validated the utility of our system in four ways: first we consider the consistency and accuracy of labels generated by non-experts using VAT second we assess the relative quality of machine learning models, or event recognizers, obtained using labels generated in VAT by non-experts, third we explore the effect of using video during labeling in VAT.

We have also built three functioning systems using event recognizers learned from labels obtained from VAT.

7.5.1 *Consistent and Accurate Labeling*

To validate the consistency of the labels obtain via VAT we conducted a user study in which participants were shown video of a person walking with a cane and data collected by the data logger. The data logger was attached to the cane. The participants were asked to label a "step" event that was broken into two pieces: (1) the part of the step where the cane tip is in contact with the ground and (2) the part of the step where the cane is not in contact with the ground. The video included 9 steps to be annotated over less than 4 min. The video and data include motion other than walking in addition to the steps to label.

For this data, the annotator agreement is high both when the annotators are compared to each other as a group and when annotators are each compared to a ground truth set of labels.

Ten participants (5 male, 5 female) were recruited from an introductory programming class. All were full-time college students. Although they were recruited from an introductory programming class we selected only students with non-STEM majors. Participants received $30 US for their participation.

We then evaluated these annotations using Krippendorff's alpha (henceforth "α") coefficient, a common statistical measure of annotator agreement (Krippendorff 2012) (We also experimented with the more limited Cohen's Kappa statistic (Cohen 1960), similar results are obtained when it is used). α is typically used in content analysis where textual units are categorized by annotators, which is very similar to the labeling of event pieces we have asked our annotators to do. α is well suited to the evaluation of raw annotations each annotation being a single label for each sample (each time step in our case). The coefficient is a single scalar value which is insensitive to the numbers of annotators, small sample sizes, unequal sample sizes and missing data. The α for our group of annotators is 0.95, which is quite high: values above 0.80 are considered reliable (Krippendorff 2012).

In addition to the study subjects, a team of three researchers annotated the data using the a-b-arbitrate (Hansen et al. 2013) approach to create a gold standard annotation. In the a-b-arbitrate approach, annotators "a" and "b" make sections and the arbitrator breaks ties if needed. In our case there was only one tie which had to be broken by the arbitrator. We then computed the pair-wise α for each study participant and the gold-standard data. The average of the pair-wise α coefficients is 0.935 with standard deviation 0.073.

7.5.2 *Quality of Machine Learning Models*

In addition to labeling the cane data, we also asked each of the 10 participants to label "scoop" data as shown earlier in Fig. 7.4. The design of the scoop counter is similar to other systems in which accelerometers attached to cooking utensils are used to detect events (Plötz et al. 2011; Pham and Olivier 2009). In this part of the study, participants were asked to collect the data, including the video, and to label scoop events in the data they had collected.

The purpose of this part of the study was to measure the quality of the event recognizers learned from data and labels generated by non-experts. Because participants each labeled their own video and data, we can not compute α to measure annotator reliability for this group.

We could have measured α for the learned recognizer but we instead use a metric more suited for assessing the recognizer as part of an interactive system. In our intended application involving construction of interactive physical objects, what matters most is identifying events quickly enough to build interactive systems. In that context, a delay of 0.5 s or less may be sufficient.

Our objective is for the trained learner to infer the completion of an event or piece of an event 0.5 s early or 0.5 s late relative to actual occurrence of the event (henceforth "β"). We use an a-b-arbitrate process to identify the actual event boundaries.

We computed an F1 metric using β by counting true positives, false negatives, and false positives as follows: An event that is inferred within 0.5 s of a true event it is a true positive, an event that is inferred by the learner near no true event or near a true event that has already been accounted for by some other, closer, inferred event is a false positive, and a true event that occurs but no inferred event is within 0.5 s of it is a false negative.

In our study of the quality of machine learned models for recognizing scoop events, we collected 12 events and annotations from each participant, used 9 as training data and 3 as test data. Event recognizers learned from the participants' data and labels achieved an average precision of 0.91 (sd = 0.15), average recall of 1 (sd = 0.0) and average F1 of 0.92 (sd = 0.93).

Figure 7.5 illustrates the sensitivity of the learner to the number of samples used as training data. The vertical axis shows the average precision, recall and F1 score for all 10 participants and the horizontal axis shows the number of labeled events used as input. For this data set, learning slows at 6 samples suggesting that the payoff associated with labeling more than 6 samples is minimal.

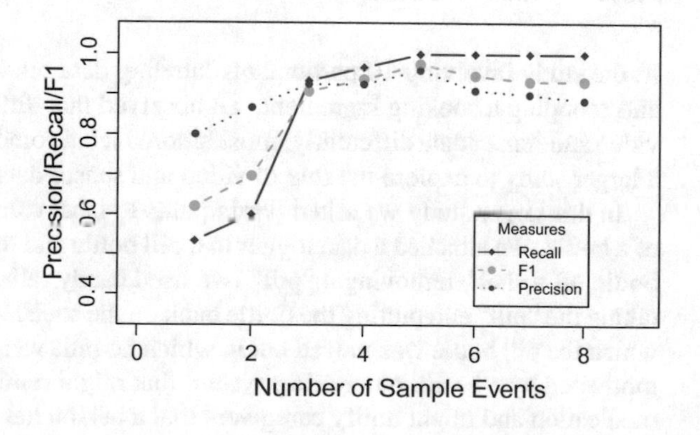

Fig. 7.5 Average learner performance as a function of the number of events labeled. Labeling more than 5 or 6 events produces little improvement in the learned recognizer. The learner was trained and evaluated on data from a single participant at a time

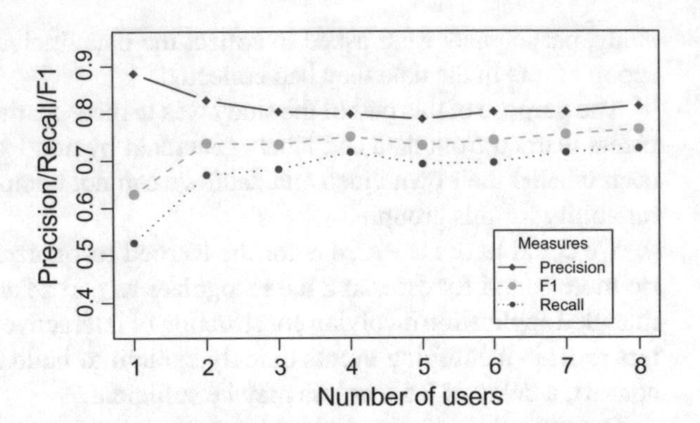

Fig. 7.6 Average learner performance on scoop events generated by three people who's data was not included in the training set as a function of the number of participants' data used in training. Learning a recognizer from data collected by 2 people results in a more general recognizer

Figure 7.5 does not show how the learned recognizer generalizes to other scoops generated by other people—the graph only shows the performance of the learner on data from a single user as labeled by that user.

Figure 7.6 shows how the learner generalizes on this data. The vertical axis in Fig. 7.6 shows the average performance of the learner on scoop data generated by three people who did not participate in the study. As before, ground truth event labels for these three samples were generated using a-b-arbitrate. The horizontal axis shows the number of participants' data used to learn the model. For example, "2" on the horizontal axis means that we combined the scoop data from 2 participants to train a recognizer and then used that recognizer on the amalgamated scoop data.

Collecting data from 2 people lead to a more general event recognizer than collecting data from 1 person, as expected. Collecting data from more than 2 people led to slight improvements in generality on this data set. These results indicate that, for scooping, a person would need to recruit two people to get a reasonably general recognizer but that recruiting more will lead to increasingly general recognizers.

7.5.3 Video Modality

In the study involving 10 participants labeling data related to walking with a cane and scooping a cooking ingredient, we observed that different participants used the video and sensor data differently; most ignored it but some relied on it. We conducted a larger study to explore the role of video and sensor data in event labeling.

In this larger study we asked participants to label data related to taking pills out of a bottle. We attached a data logger to a pill bottle and recorded a person taking the bottle off a shelf, removing a "pill" (we used candy rather than actual medication), taking the "pill" and putting the bottle back on the shelf. We included some events in which the pill bottle was moved but in which no pills were taken. This application is motivated by a health monitoring system that might remind a person to take a daily medication and might notify caregivers that a person has taken their medication.

Sensor data collected by the data logger for this example excludes flat regions in which sensor values are constant over time when the pill bottle is stationary between pill taking events. Most participants could easily identify the beginning and end of the pill taking events without the video. However, identifying pieces of events without the video proved more difficult.

The VAT interface presents two kinds of information to the user for use in labeling as shown earlier in Fig. 7.3. The two kinds of information for labeling are data and video. The data is the output of the data logger plotted against time and the video is the video captured by the camera.

We showed participants different information during labeling in order to understand the impact of the different elements on labeling. Each participant completed two labeling sessions during their visit. In each session they labeled a different collection of 5 pill-taking events and 2 just-move events. In each session they were shown just the data, just the video or both. To minimize the impact of a learning effect, we counterbalanced the order in which participants saw information for labeling. The participants were divided into 4 groups as follows:

1. 15 participants saw only data followed by both data and video,
2. 15 participants saw video and data followed by only data,
3. 15 participants saw only video followed by both data and video, and
3. 15 participants saw both data and video followed by only video.

We did not compare labeling using only video to labeling using only data because we anticipated that labeling using only video would be similar to labeling using both video and data. Our results support that anticipation. We measured the time taken to complete the labeling takes and calculated Kripendorff's α (Krippendorff 2012) for the participants' labels as described above. We also administered a short survey after the labeling session.

We examined the annotation using two-way ANOVA to look for affects involving the interface modality (users using video only, data only or both) and to check for a learning affect. The interface mode has a significant effect (Pr(>F) = 4.2e–06), but neither the learning effect (Pr(>F) = 0.28) nor the interaction of modality and learning had (Pr(>F) = 0.57) have a significant affect.

Since the affect of user interface modality was statistically significant in the ANOVA, we also looked at this affect using various two-tailed t-tests as indicated below to further explore this affect as shown in Table 7.1, which shows the value of the t-statistic, the degrees of freedom used in the test and the resultant p-value. Including video in the VAT interface always leads to significant differences in annotator performance compared to including only sensor data in the interface.

When we consider the time taken by the participants there is a significant learning effect, but only for those that used data and both video and data together (in either order) as shown in Table 7.2

Note that the mean value for data only is below the level where even tentative conclusions can be drawn. The inclusion of video ("Both") brings the average α to 0.79 (Table 7.3), which is well into the range where tentative conclusions can

Table 7.1 Including video in the VAT interface has significant impact on annotator accuracy. Tests marked with (p) are paired (unpaired tests are Welch). Bold rows are significant at the Bonferroni corrected level of **0.05/4**

	t	df	p-value
Video only or Both versus Data only	**4.00**	**35.2**	**0.0003**
Data only versus Both (p)	**4.79**	**29**	**4.58e–05**
Video only versus Both (p)	1.03	29	0.310
Video only versus Data only	**3.32**	**46.1**	**0.0017**

Table 7.2 T-tests on time taken to perform labeling involving interface modality. Bold row is significant at the **0.05/3** Bonferroni corrected level

	t	df	p-value
Both first versus Both second	1.44	59	0.16
Data then Both versus Both then Data	**2.56**	**29**	**0.016**
Video then Both versus Both then V.	−0.25	29	0.80

Table 7.3 Interface modality means and standard deviations

	Mean	Std Dev
Video only and Both	0.79	0.12
Data only	0.62	0.22
Video only	0.77	0.13
Both	0.79	0.12

be made. The point where reliable conclusion can be drawn is 0.80 (Krippendorff 2012).

Recall that the purpose of our approach is to obtain reliable learned models for recognizing certain events. In the context of this application, we seek to identify when the person has taken the pill. Finding the end of the event, that is, the point when the bottle goes from being in motion to a rest state on the shelf is easy. However in this application we would like to be able to differentiate taking a pill from merely moving the bottle around. Furthermore, we would like to identify the taking of the pill with minimum latency. Thus the most important point in time for us to identify in this application is the moment when the lid is back on, and the when we believe that the bottle is on its way back to the shelf; this is the first moment that we can believe that the entire event of pill taking is complete. For this reason we also evaluated our β metric obtained from this point in time.

As with the α data, we use two-way anova to looked for affects involving the interface modality and to check for a learning affect. As before, the interface mode has a significant effect (Pr(>F) = 1.27e–06), there is potently a learning effect

Table 7.4 Interface modality t-tests on model accuracy on a held-out test set. Tests marked with (p) are paired (unpaired tests are Welch). Bold rows are significant at the Bonferroni corrected level level of **0.05/4**

	t	df	p-value
Video only or Both versus Data only	**4.76**	**45.75**	**1.97e–05**
Data only versus Both (p)	**5.58**	**29**	**5.10e–06**
Video only versus Both (p)	1.57	29	0.128
Video only versus Data only	**2.62**	**57.57**	**0.01127**

(Pr(>F) = 0.08), but no evidence of interaction effects (Pr(>F) = 0.20). We looked for a learning affect using t-tests structured in various ways but found none.

Since the affect of user interface modality was statistically significant in the ANOVA, we again looked at this affect using various two-tailed t-tests as indicated below to further explore this affect as shown in Table 7.4.

As was the case with the annotation metric α, our results with β were not just statistically significant, they were also of practical significance. With just 5 training examples of this more complex task we obtain an average 0.21 better F1 scores on the test data using the interface with both video and data (relative to just data alone) and 0.15 better training on video alone.

After participants completed both labeling tasks, we asked them if the data or the video were more intuitive to use as part of the labeling task. Of the 30 participants who labeled using the data or both data and video (in either order), 25 said that the video was more intuitive.

7.5.4 Building Interactive Systems

We have implemented three systems as case studies: a cooking scoop counter, a medication monitor and a bicycle turn signal recognizer. In each of these, we use an event recognizer learned from data labeled in VAT along with a program that invokes the recognizer and responds to events. This program also provides the mapping between events and callback functions to the recognizer, as shown in Fig. 7.7. The recognizer uses this mapping to call the appropriate function when an event is detected. The recognizer runs repeatedly on the data it receives from the wirelessly-connected data logger, and spawns a thread for each callback when events are detected.

Cooking Scoop Counter

Using the data recorded by the ten scoop study participants discussed earlier, we trained a recognizer that recognizes when the spoon is used to scoop material out of a canister. On each scoop event, a count is incremented and announced out loud.

```
def left_turn(event):
    leds.left_turn_on(8)

def right_turn(event):
    leds.right_turn_on(8)

event_hooks = {
    'Right Turn': right_turn,
    'Left Turn': left_turn,
}

RealTimeRecognizer(expanduser('~/bike/model.pkl'),
    event_hooks=event_hooks).run()
```

Fig. 7.7 A shortened example of the code for the hand signals

Medication Reminder

This system turns on a sign reminding a patient to take their pills and waits for a smart pill bottle to notice if it has been opened and closed (based on movement data), or only picked up and moved. On detecting the event, the application turns off the sign and sends a text message to the caretaker. To build this system, we attached a data logger to the side of an empty pill bottle, filled the bottle with candy, and collected sample data with three different users.

In order to understand the amount of time required to collect and label data for the medication reminder, we recruited 4 university students with design backgrounds (3 male, 1 female) to collect the data required to train the event recognizer while recording the time required to complete each step. Participants each required between 59 and 80 min, with an average of 71 min, to collect and label the video and data. A little more than half of that time, 39 min, was spent labeling scoop events in the data.

Bicycle Hand Signal

Using VAT, we made a bike riding accessory that interprets a biker's hand signals and displays the corresponding turn signal on the rider's back. Data and video were captured outdoors in a park using a quadcopter to record the video. The quadcopter was flown behind the rider to allow the video camera to capture hand signals made by the rider. A still frame from the video is shown in Fig. 7.3 (the quadcopter shadow is above the rider's left arm). The data logger was attached to a glove worn on the rider's left hand as shown in Fig. 7.2.

The turn signal event recognizer was learned from 20 labeled events including left turns, right turns, and stopping. The LEDs and a laptop are mounted in a backpack that the rider wears. The rider can make normal biking hand signals while wearing the glove, and the application will automatically detect and display the matching signals on the LED. Figure 7.7 shows an example of the code used to integrate with the recognizer, note that the complexity of the learning algorithm is entirely hidden.

7.6 Conclusion

Adding synchronized video to sensor data for labeling creates a more intuitive labeling experience and supports increased precision in labeling. This approach allows designers to quickly create machine learning models for physical interactive devices which must recognize events in the physical world. We have run several user studies to document behaviour of non-expert users in the context of annotating physical event and to demonstrate the advantages of our approach. More importantly, we demonstrated the value of this approach by building three example systems using event recognizers learned from labeled examples using our process: a cooking scoop counter, a medication monitor and a gesture-based bike turn signal system.

Future work might include other means for supporting the annotation process, for example pre-annotation. In addition the learning algorithm could be replaced or modified to improve the efficiency and performance of the learner.

Acknowledgements This work supported by NSF grant IIS-1406578.

References

Bulling A, Blanke U, Schiele B (2014) A tutorial on human activity recognition using body-worn inertial sensors. ACM Comput Surv (CSUR) 46(3):33

Chen SS, Gopalakrishnan PS (1998) Clustering via the Bayesian information criterion with applications in speech recognition. In: Proceedings of the 1998 IEEE international conference on acoustics, speech and signal processing, 1998, vol 2. IEEE, pp 645–648

Cohen J (1960) A coefficient of agreement for nominal scales. Educ Psychol Meas 20:37–46

Hansen DL, Schone PJ, Corey D, Reid M, Gehring J (2013) Quality control mechanisms for crowdsourcing: peer review, arbitration, & expertise at familysearch indexing. In: Proceedings of the 2013 conference on computer supported cooperative work. ACM, pp 649–660

Harrison C, Hudson SE (2009) Providing dynamically changeable physical buttons on a visual display. In: Proceedings of the SIGCHI conference on human factors in computing systems, CHI '09. ACM, New York, pp 299–308. https://doi.org/10.1145/1518701.1518749

Hartmann B, Klemmer SR, Bernstein M, Abdulla L, Burr B, Robinson-Mosher A, Gee J (2006) Reflective physical prototyping through integrated design, test, and analysis. In: Proceedings of the 19th annual ACM symposium on User interface software and technology, UIST '06. ACM, New York, pp 299–308. https://doi.org/10.1145/1166253.1166300

Hartmann B, Abdulla L, Mittal M, Klemmer SR (2007) Authoring sensor-based interactions by demonstration with direct manipulation and pattern recognition. In: Proceedings of the SIGCHI conference on human factors in computing systems, CHI '07. ACM, New York, pp 145–154 (2007). https://doi.org/10.1145/1240624.1240646

Hudson SE, Mankoff J (2006) Rapid construction of functioning physical interfaces from cardboard, thumbtacks, tin foil and masking tape. In: Proceedings of the 19th annual ACM symposium on user interface software and technology, UIST '06. ACM, New York, pp 289–298. https://doi.org/10.1145/1166253.1166299

Jin H, Xu C, Lyons K (2015) Corona: positioning adjacent device with asymmetric bluetooth low energy RSSI distributions. In: Proceedings of the 28th annual ACM symposium on user interface software & technology, UIST '15. ACM, New York, pp 175–179

Jones M, Walker C, Anderson Z, Thatcher L (2016) Automatic detection of alpine ski turns in sensor data. In: Proceedings of the 2016 ACM international joint conference on pervasive and ubiquitous computing: adjunct, UbiComp '16. ACM, New York, pp 856–860. https://doi.org/10.1145/2968219.2968535

Jones MD, Johnson N, Seppi K, Thatcher L (2018) Understanding how non-experts collect and annotate activity data. In: Proceedings of the 2018 ACM international joint conference and 2018 international symposium on pervasive and ubiquitous computing and wearable computers, UbiComp '18. ACM, New York, pp 1424–1433. https://doi.org/10.1145/3267305.3267507

Jurafsky D, Martin J (2009) Speech and language processing: an introduction to natural language processing, computational linguistics, and speech recognition. In: Prentice Hall series in artificial intelligence. Pearson Prentice Hall

Krippendorff K (2012) Content analysis: an introduction to its methodology. Sage

Laput G, Yang C, Xiao R, Sample A, Harrison C (2015) EM-Sense: touch recognition of uninstrumented, electrical and electromechanical objects. In: Proceedings of the 28th annual ACM symposium on user interface software & technology, UIST '15. ACM, New York, pp 157–166. https://doi.org/10.1145/2807442.2807481

Murphy KP (2012) Machine learning: a probabilistic perspective. In: Adaptive computation and machine learning series. MIT Press

Patterson DJ, Fox D, Kautz H, Philipose M (2005) Fine-grained activity recognition by aggregating abstract object usage. In: Proceedings of the ninth IEEE international symposium on wearable computers, 2005. IEEE, pp 44–51

Pham C, Olivier P (2009) Ambient intelligence: European conference, AmI 2009, Salzburg, Austria, November 18–21, 2009. Proceedings, chap. Slice&Dice: recognizing food preparation activities using embedded accelerometers. Springer, Berlin, Heidelberg, pp 34–43. https://doi.org/10.1007/978-3-642-05408-2_4

Plötz T, Moynihan P, Pham C, Olivier P (2011) Activity recognition and healthier food preparation. In: Activity recognition in pervasive intelligent environments. Springer, pp 313–329

Sakoe H, Chiba S (1978) Dynamic programming algorithm optimization for spoken word recognition. IEEE Trans Acoust Speech Signal Process 26(1):43–49. https://doi.org/10.1109/TASSP.1978.1163055

Savage V, Follmer S, Li J, Hartmann B (2015a) Makers' marks: physical markup for designing and fabricating functional objects. In: Proceedings of the 28th annual ACM symposium on user interface software & technology. ACM, pp 103–108

Savage V, Head A, Hartmann B, Goldman DB, Mysore G, Li W (2015b) Lamello: passive acoustic sensing for tangible input components. In: Proceedings of the 33rd annual ACM conference on human factors in computing systems, CHI '15. ACM, New York, pp 1277–1280

Young S, Evermann G, Gales M, Hain T, Kershaw D, Liu X, Moore G, Odell J, Ollason D, Povey D et al (1997) The HTK book, vol 2. Entropic Cambridge Research Laboratory Cambridge

Zhang Y, Harrison C (2015) Tomo: wearable, low-cost electrical impedance tomography for hand gesture recognition. In: Proceedings of the 28th annual ACM symposium on user interface software & technology, UIST '15. ACM, New York, pp 167–173. https://doi.org/10.1145/2807442.2807480

Chapter 8
A Multi-media Exchange Format for Time-Series Dataset Curation

Philipp M. Scholl, Benjamin Völker, Bernd Becker and Kristof Van Laerhoven

Abstract Exchanging data as character-separated values (CSV) is slow, cumbersome and error-prone. Especially for time-series data, which is common in Activity Recognition, synchronizing several independently recorded sensors is challenging. Adding second level evidence, like video recordings from multiple angles and time-coded annotations, further complicates the matter of curating such data. A possible alternative is to make use of standardized multi-media formats. Sensor data can be encoded in audio format, and time-coded information, like annotations, as subtitles. Video data can be added easily. All this media can be merged into a single container file, which makes the issue of synchronization explicit. The incurred performance overhead by this encoding is shown to be negligible and compression can be applied to optimize storage and transmission overhead.

8.1 Introduction

At the heart of each Activity Recognition task is a dataset. This dataset might be formed from multiple media streams, like video, audio, motion and other sensor data. Recorded at different rates, sparsely or uniformly sampled and with different numerical ranges, these streams are challenging to process and store. Commonly, datasets are published in multiple character-separated values (CSV) files, either with a constant rate or time-coded. For small, independent time-series this is a worthwhile approach, mostly due to its simplicity and universality. However, when observing with multiple independent sensors, synchronization quickly becomes a challenge.

P. M. Scholl (✉) · B. Völker · B. Becker
University of Freiburg, Freiburg, Germany
e-mail: pscholl@informatik.uni-freiburg.de

D. Völker
e-mail: voelkerb@informatik.uni-freiburg.de

K. V. Laerhoven
University of Siegen, Siegen, Germany
e-mail: kvl@eti.uni-siegen.de

N. Kawaguchi et al. (eds.), *Human Activity Sensing*,
Springer Series in Adaptive Environments,
https://doi.org/10.1007/978-3-030-13001-5_8

Different rate recordings have to be resampled, time-coded files have to be merged. Storing such data in several (time-coded) CSV files hides this issue, until the dataset is going to be used. Furthermore parsing CSV files incurs a large performance and storage overhead, compared to a binary format.

Examples of Activity Recognition Datasets

HASC Challenge (Kawaguchi 2011): >100 subjects, time-coded CSV files.
Box/Place-Lab (Intille et al. 2006): A sensor-rich home, in which subjects are monitored for long terms. Data is available in time-coded CSV files.
Opportunity (Roggen et al. 2010): 12 subjects were recorded with 72 on- and off-body sensors in an Activities of Daily Living (ADL) setting. Multiple video cameras were used for post-hoc annotations. Data is published in synchronized CSV files.
Kasteren's Home (Kasteren et al. 2010): 12 sensors in 3 houses. Data is stored in matlab files.
Borazio's Sleep (Borazio et al. 2014): 1 sensor, 42 subjects. Data is stored in numpy's native format.
Freiburg Longitudinal (van Laerhoven 2019): 1 sensor, 1 subject, 4 weeks of continuous recording. Data is stored in numpy's native format.

One alternative is storing such datasets in databases, like SQL, NOSQL or HDF5-formats. This eases the management of large datasets, but shows the same issues as a CSV format, namely that there is no direct support for time-series or video data. An alternative approach is to store time-series in existing multi-media formats. Encoding all multi-media data in one file allows to merge streams, to synchronize them and to store (meta-)data in a standardized format. In the next section, we will first look at the formats commonly used to exchange data in Activity Recognition, afterwards we detail a multi-media format and evaluate the incurred performance and storage overhead of each format.

8.2 Related Work

In the classic activity recognition pipeline (Bulling et al. 2014), the first step is to record and store sensor data. The observed activities, executed by humans, animals or other actors, are recorded with different sensors. Each sensor generates a data *stream*, whether this is a scene camera for annotation purposes, a body-worn motion capturing system or binary sensors like switches. Sampled at different rates, with differing resolutions, ranges, units and formats these streams offer a large variety of recording parameters. These parameters are usually documented in an additional file that resides next to the actual data The actual data is commonly stored in a CSV file, in a binary format for Matlab or NumPy, or in format specific to some Machine Learning framework like ARFF (Hall et al. 2009) or libSVM (Chang et al. 2011).

Synchronizing such multi-modal data, i.e. converting this data to the same rate and making sure that recorded events happened at the same time presents a major challenge Possible approaches range from offline recording with post-hoc synchronization on a global clock, to live streaming with a minimum delay assumption—all but the last one require some form of clock synchronization and careful preparation. Storing events with timestamps on a global clock is then one possible way to allow for post-recording synchronization, i.e. each event is stored as a tuple of `<timestamp, event data>`.

The subsequent step of merging such time-coded streams often requires to adapt their respective rates. Imagine, for example, a concurrent recording of GPS at 3 Hz and acceleration at 100 Hz. To merge both streams: will GPS be upsampled or acceleration downsampled, or both resampled to a common rate? Which strategy is used for this interpolation, is data simply repeated or can we assume some kind of dependency between samples? How is jitter and missing data handled? These question need to be answered whenever *time-coded* sensor data is used. A file format which makes the choice of possible solutions explicit is the goal of this paper.

8.3 Multi-media Container Approach

Sensor data commonly used in Activity Recognition is not different from low-rate audio or video data. Common parameters are shared, and one-dimensional sensor data can be encoded with a lossless audio codec for compression. Rate, sample format and number of channels need to be specified for an audio track. The number of channels is equivalent to the number of axis an inertial sensor provides, as well as its sample rate. The sample format, i.e. how many bits are used to encode one measurement, is also required for such a sensor. Other typical parameters, like the range settings or conversion factor to SI units (if not encoded as such), can be stored as additional meta-data, as those are usually not required for an audio track.

Lossless compression, like FLAC (2019) or WavPack (2019), can be applied to such encoded data streams. This allows to trade additional processing for efficient storage. In the evaluation section several lossless schemes are compared. These include the general LZMA2 and ZIP compressors, and the FLAC (2019) and WavPack (2019) audio compressors. All but the first two can be easily included in multi-media container formats. To use audio streams, data needs to be sampled at a constant rate, i.e. the time between two consecutive samples is constant and only jitter smaller than this span is allowed. Put differently, the time between two consecutive data samples t_i and t_{i+1} at frame i must always be less than or equal to the sampling rate r: $\forall i \in N : t_{i+1} - t_i \leq \frac{1}{r}$. Compared to time-coded storage, the recording system has be designed to satisfy this constraint. Problems with a falsely assumed constant rate recording setup will therefore surface faster. Especially in distributed recording settings, where the just mentioned constraint is checked only against local clocks which might drift away from a global clock, is a common challenge.

Sparsely sampled events can be encoded as subtitles. Here, each sample is recorded independently of its preceding event, i.e. the above mentioned constraint does not hold. Each event needs to be stored with a time-code and the actual event data. Depending on the chosen format, this can also include a position in the frame of an adjacent video stream or other information. For example, this can be used to annotate objects in a video stream. A popular format is the Substation Alpha Subtitle (SSA 2019) encoding, which includes the just mentioned features. Since data is encoded as strings, it is suitable for encoding ground truth labels. To a limited extent, since no compression is available, it can be used for sensor events as well. For example, low rate binary sensors, like RFID readers could be encoded as subtitles.

Encoded sensor and subtitle data can then be combined with audio and video streams in a multi-media container format. One such standard is the Matroska (2016) format, that is also available in a downgraded version called WebM (2019) for web-browsers. Once the data streams are combined into one such file, this data can be "played" back in a synchronous manner. This means that streams recorded at different rates, and in different formats, need to be converted to a common rate and possibly common format. Meta-data that contains additional information like recording settings, descriptions and identifiers can be stored in addition to the parameters already contained in the stream encoding. For this task off-the-shelf software, like FFmpeg (2019) can be used, which also provides functionality like compression, resampling, format conversion and filtering. Annotation tasks can be executed with standard subtitle editing software, discouraging the creation of yet another annotation tool. Furthermore, video streaming servers can be used for transporting live sensor data recordings to remote places.

The use of such a standard format for curating datasets allows for re-using existing software, however not without limitations. Asynchronous, also called sparsely sampled, data recorded at high rates is not supported. This mainly stems from the simplifying assumption that streams are recorded with a constant rate. We presume that satisfying this constraint while recording to be easier than handling asynchronicity later on. For example, breaks, shifts or jitter due to firmware bugs can be detected earlier. In general this is a hard limitation, however different data types can also be encoded in multiple streams. Also, the en- and decoding overhead might be a limitation, which we will look at in the next section.

Compared to the de-facto standard of using CSV files, encoding sensor data as audio, annotations as subtitles and audio- and video-data in standard formats provides several improvements. Important parameters like sampling rate, format and number of axes are included in the file. Adding additional information as meta-data leads to a completly *self-descriptive* format. *Synchronous* playback of multiple streams, which requires re-sampling, is supported by off-the-shelf software. Related problems, like un-synchronized streams can be caught earlier, since this step is explicit. The container format is *flexible* enough to support different number formats, i.e. values can be encoded as floats or integers of varying bit-size. Optional compression leads to *compact* storage, which allows for efficient storage and transmission. Additionally, when thinking about large datasets, such a container format requires *divisible* storage. This functionality (seeking without reading the whole dataset into memory[1]) is provided. Such divisible storage also allows for streaming applications, which, for multi-media format, also provides network protocols to cover transmission via un-reliable links.

8.4 Evaluation

Compressing sensor data as an audio stream incurs an en- and decoding overhead, and provides optimized storage. In this section both are quantified. By a repetitive measurement of the relative wall clock time for decompression, the processing overhead is measured. This runtime overhead is reported as the fraction of reading time relative to reading and *converting* the CSV file into main memory. The compression factor is determined by comparing the number of bytes required to store the compressed file to the original, deflated CSV file. Binary and text-based storage is compared. The Zip and LZMA2 algorithms are used for general byte-wise compression, and the lossless FLAC and WavPack compressor for audio-based compression. Additionally storing in the sqlite3, and MongoDB database, as well as the HDF5 format is compared. The approach of compressing binary files with a general compressor is used by Numpy, MongoDB and HDF5 for example.

The test were run on the Kitchen CMU (De la Torre et al. 2009), Opportunity (Roggen et al. 2010), HASC Challenge (Kawaguchi 2011) and on twenty days of the Freiburg Longitudinal Wrist (van Laerhoven 2019) datasets. A machine with an i7-4600U CPU running at 2.1 GHZ with 8 GB of memory was used for all tests. Figure 8.1a, b show the results of these tests, *csv/gz* refers to a zip-compressed CSV file, *csv/bz2* to an LZMA2 compressed file,[2] *f32* refers to a 32-bit-float binary format, *wv32* to WavPack compression of 32-bit-floats, and *flac* to the *FLAC* compressor which only supports 24 bits values. *nosql* refers to storing the data in a MongoDB (*NoSQL*) database, *sql* to sqlite3 storage, and *hdf5/gz* to encoding the sensor data in a zip-compressed HDF5 container. For *MongoDB* storage, each data stream is stored together with its metadata into a so called *collection* inside the database. Since MongoDB's *BSON* format only allows for certain data types, we choose the 64 bits *Double* format to store the individual event data with the default compression parameters. Equivalent to the *MongoDB* structure, each datastream is stored as a separate table in the *sqlite* database and the event data is stored in 64 bits *REAL* format. The stream's metadata is stored in a separate table. De- and encoding was performed using the corresponding python interfaces *pymongo* and *sqlite3*. The *HDF5* Group (2019) data was generated and de-/encoded using the *h5py* python interface and stored with zip-compression.

[1]Which would be required for time-coded storage.

[2]The XZ utils package was used.

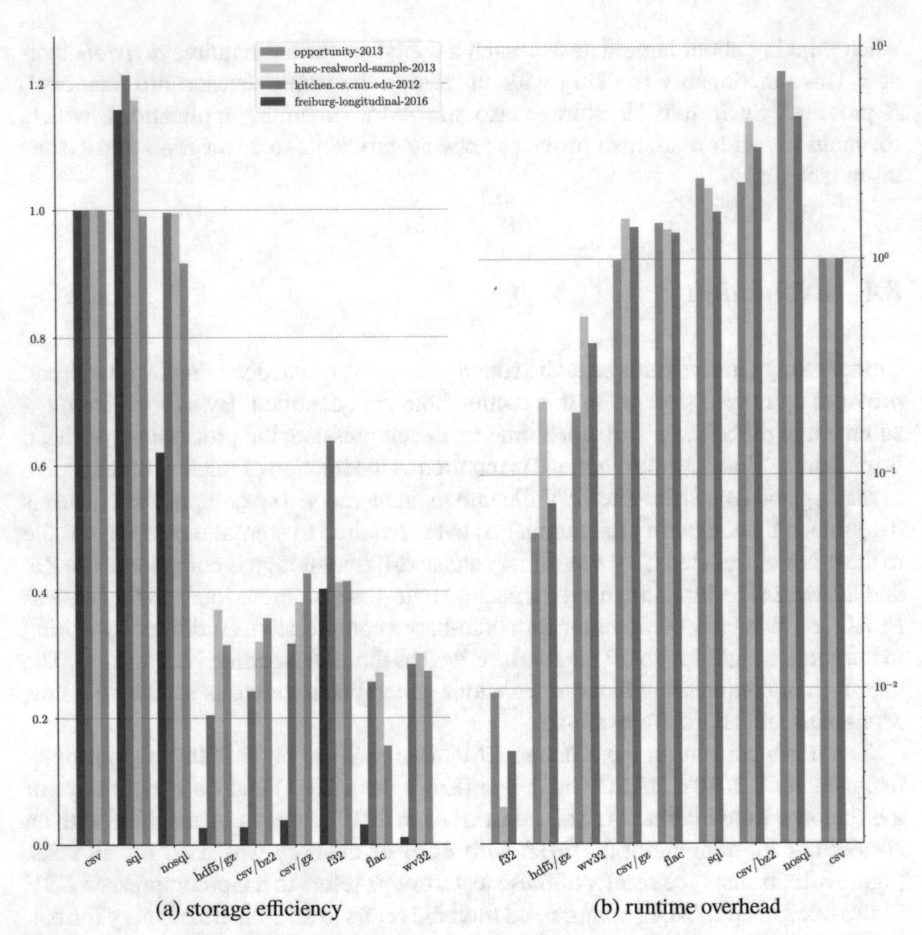

Fig. 8.1 Comparison of runtime overhead and storage efficiency, which is to be traded off for each format. Each is given relative to a deflated CSV file

8.4.1 Processing Overhead

It is interesting to check how much overhead is incurred for decompression by each storage paradigm, as this gives an insight if data needs to be stored in an intermediate format while evaluating a recognition pipeline. If the overhead is comparatively low, no intermediate storage format is required and data can always be loaded from such a file. However, should decoding require more processing time than the actual processing, an intermediate file needs to be used.

Naturally such a format would be binary, at best a memory image which can be mapped into main memory as a 32 bit-float. The baseline is, therefore, the time required to convert a CSV from disk into a binary format in memory. The fraction of time required to do the same for each storage scheme is reported in Fig. 8.1b. Each test is repeated six times, and the first run is discarded, i.e. data is always read from the disk cache.

Just parsing a CSV file incurs an up to hundred-fold overhead compared to reading a binary file (*f32* in Fig. 8.1b). Compressing CSV data[3] can increase the runtime by 1.4–3.0 times. So, looking only at runtime performance a CSV or compressed CSV file should hardly be used for large datasets. When comparing compression schemes, it can be seen that a 32 bit WavPack compression provides the second lowest runtime overhead, only the hdf5-scheme is roughly twice as fast. Storing data into a *MongoDB* database comes with the cost that interfacing is done over a TCP socket in plaintext JSON format, which incurs an up to 4-times overhead compared to CSV, and at least 100-fold compared to raw binary storage. A trade-off between storage efficiency and performance has to be found.

8.4.2 Storage Efficiency

General compression and audio compression algorithms were tested. Raw binary, WavPack (2019) and FLAC (2019) compression were taken from the FFmpeg (2019) suite with default parameters. Figure 8.1a shows the amount of compression that was achieved for each dataset per algorithm compared to uncompressed CSV files.

The used datasets show different characteristics found in other datasets as well. For example the Longitudinal (van Laerhoven 2019) dataset can be massively compressed with general algorithms, almost down to 2% of its original size. This is mainly owed to the fact that the contained acceleration data was recorded with a resolution of only 8-bits, and that a run-length compression was already applied during recording. This run-length compression is deflated for CSV storage first, adding a lot of redundancy. Generally, text formats provide a larger storage efficiency only when less characters are required per sample than their binary counterparts.

This effect is partially visible for the kitchen dataset De la Torre et al. (2009). The relative storage requirements for binary storage (f32 in Fig. 8.1a) is a lot larger than for other datasets. Here, the text representation of values is smaller since they range from 0 to 1, i.e. the text values are almost always of the same length. Other datasets provide a range of values with larger pre-decimals, hence a longer text representation. The maximum dynamic range that can be stored with a text-based format more efficiently is therefore limited to the (decimal) encoding, (less than 10,000 for five digits), while a comparable binary encoding can range up to 2^{5*8}.

When optimizing data for space efficiency, the encoding of each value is the most critical factor. Limiting the number of bits per value, in essence assuming a limited dynamic range of the encoded signal, has the strongest influence on the storage efficiency. The 24 bit flac encoding shows the best overall efficiency due to this. If a dynamic range of more than 2^{24} is required, the wavpack encoding should be

[3]Note that the 8.1b represent the factor between the compression and simply *reading* an uncompressed CSV file.

considered. However, when encoding values in text format with a limited dynamic range (equivalent to four characters), a text compression algorithm is not worse than encoding data in binary format. For the general case and when binary storage can be used, the WavPack compression provides the same storage efficiency as the more general LZMA2 compressor and provides comparable compression factors to the hdf5 format.

NoSql, and SQL storage do hardly provide a benefit over CSV concerning storage efficiency in our tests. For SQL, this is probably an implementation artifact, since sqlite3 stores date internally as a string representation. NoSql, represented via MongoDB here, stores data in a compressed binary JSON format. This provides a benefit only for one dataset, which contains a lot of redundancy.

8.5 Conclusion

Curated time-series data provides the basis for comparing Activity Recognition and other Machine Learning approaches in an objective and repeatable manner. This data usually includes low-rate sensor data, e.g. motion sensors, time-coded annotations, and second-level evidence like video and audio recordings. The state of the art for exchanging this data seems to be a mix of time-coded CSV format, and dedicated audio- and video-codecs. Synchronizing the stored data-streams is usually not done by the dataset provider and the dataset consumer is left with this challenge that usually requires information of the recording setup. This is especially problematic when video or audio data is recorded in addition to sensor data. The HDF5 which can provide a smaller decoding overhead, provides no support for decoding audio- and video-data directly.

The CSV format incurs a large overhead both in runtime and storage. A possible alternative, with lower overhead, is presented here. Motion and other sensor data, as well as extracted features, can be stored in lossless audio formats. Ground truth labels and other time-coded information can be stored in subtitle format. These streams can then be merged in a common multi-media container (e.g. Matroska), with additional video streams. One recording session is stored in a single file, that can be *self-descriptive*, *synchronized*, with a fitting *storage-runtime trade-off* for multi-media data and supports *live-streaming*.

References

Borazio M, Berlin E, Kücükyildiz N, Scholl PM, Van Laerhoven K (2014) Towards benchmarked sleep detection with inertial wrist-worn sensing units. In: IEEE international conference on health-care informatics (2014)

Bryant D. The WavPack Codec. https://www.wavpack.org/

Bulling A, Blanke U, Schiele B (2014) A tutorial on human activity recognition using body-worn inertial sensors. ACM Comput Surv 46(3):1–33. http://dl.acm.org/citation.cfm?doid=2578702.2499621

Chang CC, Lin CJ (2011) LIBSVM: a library for support vector machines. ACM Trans Intell Syst Technol 2(3):1–27

Group F. FFmpeg. https://www.ffmpeg.org/

Group TH. High-performance data management and storage suite. https://www.hdfgroup.org/solutions/hdf5/

Hall M, Frank E, Holmes G, Pfahringer B, Reutemann P, Witten IH (2009) The WEKA data mining software. SIGKDD Explor Newsl 11(1):10. http://portal.acm.org/citation.cfm?doid=1656274.1656278

Intille SS, Larson K, Tapia EM, Beaudin JS, Kaushik P, Nawyn J, Rockinson R (2006) Using a live-in laboratory for ubiquitous computing research. In: International conference on pervasive computing. Springer, Berlin, Heidelberg, pp 349–365

Kasteren TLMV, Englebienne G, Kr BJA (2010) Human activity recognition from wireless sensor network data: benchmark and software. In: Activity recognition in pervasive intelligent environments

Kawaguchi N, Ogawa N, Iwasaki Y (2011) Hasc challenge: gathering large scale human activity corpus for the real-world activity understandings. In: Proceedings of the 2nd augmented human international conference, p 27. http://dl.acm.org/citation.cfm?id=1959853

van Laerhoven K. Longitudinal wrist motion dataset. https://earth.informatik.uni-freiburg.de/application/files/hhg_logs/0089/index.html

Lamparter D. Advanced sub station alpha. http://fileformats.wikia.com/wiki/SubStation_Alpha

Matroska NPO (2016) The Matroska file format. https://www.matroska.org/

Roggen D, Calatroni A, Rossi M, Holleczek T, Kilian F, Tr G, Lukowicz P, Bannach D, Pirkl G, Ferscha A, Doppler J, Holzmann C, Kurz M, Holl G, Creatura M, Mill R (2010) Collecting complex activity datasets in highly rich networked sensor environments. In: International conference on information processing in sensor networks

De la Torre F, Hodgins J, Montano J, Valcarcel S, Forcada R, Macey J (2009) Guide to the Carnegie Mellon University multimodal activity (CMU-MMAC) database. Carnegie Mellon University, Robotics Institute

WebM C. The WebM file format. http://www.webmproject.org/

Xiph.org F. The Free Lossless Audio Codec (FLAC). https://www.xiph.org/flac/

Chapter 9
OpenHAR: A Matlab Toolbox for Easy Access to Publicly Open Human Activity Data Sets—Introduction and Experimental Results

Pekka Siirtola, Heli Koskimäki and Juha Röning

Abstract OpenHAR is a toolbox for Matlab to combine and unify 3D accelerometer data of ten publicly open data sets. This chapter introduces OpenHAR and provides initial experimental results based on it. Moreover, OpenHAR provides an easy access to these data sets by providing them in the same format, and in addition, units, measurement range, sampling rates, labels, and body position IDs are unified. Moreover, data sets have been visually inspected to fix visible errors, such as sensor in wrong orientation. For Matlab users OpenHAR provides code which user can use to easily select only desired parts of this data. This chapter also introduces OpenHAR to users without Matlab. For them, the whole OpenHAR data is provided as a one .txt-file. Altogether, OpenHAR contains over 280 h of accelerometer data from 211 study subjects performing 17 daily human activities and wearing sensors in 14 different body positions. This chapter shown the first experimental results based on OpenHAR data. The experiment was done using three classifiers: linear discriminant analysis (LDA), quadratic discriminant analysis (QDA), and classification and regression tree (CART). The experiment showed that using LDA and QDA classifiers and OpenHAR data, as high recognition rates can be achieved in a previously unseen test data than by using a data set specially collected for this purpose. With CART the results obtained using OpenHAR data were slightly lower.

P. Siirtola (✉) · H. Koskimäki · J. Röning
Biomimetics and Intelligent Systems Group, University of Oulu, FI-90014 Oulu, Finland
e-mail: pekka.siirtola@oulu.fi

H. Koskimäki
e-mail: heli.koskimaki@oulu.fi

J. Röning
e-mail: jjr@oulu.fi

N. Kawaguchi et al. (eds.), *Human Activity Sensing*,
Springer Series in Adaptive Environments,
https://doi.org/10.1007/978-3-030-13001-5_9

9.1 Introduction

Human activity recognition can be used to several real-life applications. Therefore, during the recent years, a lot of research have been done to recognize human activities based on inertial sensor data from wearable and smartphone sensors. However, mostly human activity recognition is used for health and fitness monitoring as many of the devices are originally designed to this purpose. Moreover, human activity recognition can be applied, for instance, for personalized advertising; smarthomes that anticipates the user's needs; self-managing system that adapts to user's activities, and for context-aware applications (Lockhart et al. 2012; Siirtola et al. 2018a).

According to Incel et al. (2013), human activity recognition process can be divided into three main phases: data collection, training and activity recognition. Therefore, the basis of a reliable recognition model is always an extensive data set from the studied problem. However, collection one is can be difficult, and especially, time consuming. Moreover, the data set does not only need to be gathered, it needs to be labeled as well. Instead of spending time to collect own data set, it is possible to train model based on publicly open data sets. In fact, it is more and more common that data sets used in the previous activity recognition studies are made publicly available. Moreover, by combining multiple publicly open data sets, bigger data set can be build and bigger data set normally means more general and accurate recognition model. However, it is not always that easy to combine data sets.

Difficult of combining data sets was noticed for instance in our previous article (Siirtola et al. 2018c) where different publicly open human activity data sets were cross-validated. The idea was to train human activity recognition models using one data set and test them using another. This way it was possible to see how well models work when data for training and testing are collected in different environments and using different sensors. As a side results, it was noticed that combining data sets is not easy. For instance, it was noted that data sets are often stored in different formats, sensor orientation varies, units are not always the same, etc.

Moreover, combining publicly open data sets is not the only problem. In fact, sometimes it is not even that straightforward to use them separately. In Siirtola et al. (2018b) personalizing human activity recognition models based on incremental learning were studied and the experiments were based on publicly open data set containing data from ten study subjects. While in the study it was noted that personalizing improves recognition accuracy, it was also noted that one subject had worn sensor in different orientation than others making this data non-uniform with other subjects data. Therefore, this persons data was not used in the experiments.

This study is an extension to Siirtola et al. (2018d), which introduced OpenHAR—a free Matlab toolbox combining publicly open data sets. It provides an easy access to accelerometer signals of ten publicly open human activity data sets. Data sets are easy to access as OpenHAR provides all the data sets in the same format, units, measurement range and labels are unified, as well as, body position IDs. Moreover, data sets with different sampling rates are unified using downsampling. What is more, data sets have been visually inspected to find visible errors, such as sensors in wrong orientation. OpenHAR improves re-usability of data sets by fixing these errors. In addition, OpenHAR provides Matlab code to easily select only desired parts of this data. The idea of combining data sets is not unique: in Bartlett et al. (2017) a data set called AcctionNet collating six publicly open data sets was introduced. This data contains over 10 million labeled accelerometer samples samples from 13 activities. The data sets used in AcctionNet are partly the same as the ones used in OpenHAR. The main difference between AcctionNet and OpenHAR is that OpenHAR is not just a data set, it also provides tools to easily select only those parts of data that are important to certain application.

This chapter introduces OpenHAR to non-Matlab users as well. For them, the whole OpenHAR data is provided as a one .txt-file. Moreover, this chapter shows the first experimental results based on OpenHAR data to show its potential. In the experiment, four human activities are classified using three classifiers (linear discriminant analysis, quadratic discriminant analysis, and classification and regression tree) to compare their performance.

The chapter is organized as follows: Sect. 9.2 introduces OpenHAR, and Sect. 9.3 explains how it can be used. Section 9.4 shows the experimental results based on OpenHAR and Sect. 9.5 contains discussion and conclusions.

9.2 OpenHAR

9.2.1 Combined Data Sets

Several human activity recognition data sets have been published as open data. The planning of OpenHAR started by surveying these. It was studied which of these are similar enough to be combined as one. Eventually, ten publicly open data sets were selected for OpenHAR, these are listed in Table 9.1. Common with these data sets is that they all contain data from the activities of daily living. In addition, we wanted to use only data sets which contain raw accelerometer data collected with reasonable sampling rate. Our survey showed that there is also other publicly open data sets available, but they were not included to this study as they do not fulfill our requirements: for instance, Reiss et al. (2012) was not included to OpenHAR as data of it filtered and not raw. Moreover, SHL data set which is an excellent and extensive activity data set by Gjoreski et al. (2018) was not included to OpenHAR as it is so huge compared to selected ten data sets that it would have a too dominant role in the combined data set.

Table 9.1 shows that there are a lot of differences between data sets. Therefore, combining these is not straightforward. File formats are are different and also data files are grouped in several different ways. As an example, in some cases, the whole data set is stored in one single file but often data are divided into multiple folders and files. In addition, when it comes to activity labels, both integers and strings were used as labels in the original data sets. Moreover, numerical labels did not have the same response in different data sets. In some cases, labels also had different meaning, for instance depending on the studied data set, activities *walking*, *walking upstairs*

Table 9.1 OpenHAR contains ten publicly open human activity data sets

Data set ID	Author	File format	Frequency (Hz)	Labels	Range and unit
1	Banos et al. (2014)	.log	50	Numeral	± 24 m/s^2
2	Ortiz et al. (2013)	.txt	50	Numeral	± 2 g
3	Shoaib et al. (2014)	.csv	50	Strings	± 20 m/s^2
4	Siirtola and Röning (2012)	.txt	40	Numeral	± 20 m/s^2
5	Stisen et al. (2015)	.csv	50–200	Strings	± 40 m/s^2
6	USC-HAD (2012)	.mat	100	Numeral	± 6 g
7	UniMib-SHAR (2017)	.mat	50	Numeral	± 20 m/s^2
8	HuGaDB (2017)	.txt	60	Numeral	± 32767
9	RealworldHAR (2016)	.csv	50	Strings	± 20 m/s^2
10	MobiAct (2016)	.csv	200	Strings	± 20 m/s^2

Table 9.2 Fixes needed for the datasets

Data set ID	Fixes
1	Timestamp added
3	subj. 8, belt: orientation fixed
4	Timestamp added
5	Sampling rates unified
6	Timestamp added
7	Timestamp added
8	Timestamp added
9	subj.8, chest: orientation fixed, subj.15, thigh: orientation fixed, subj.3, upper arm: orientation fixed, subj.3, waist: orientation fixed
10	–

and *walking downstairs* had own labels but it was also possible that label *walking* included walking at flat level and walking at stairs. Another difference in the data sets is the used sampling frequency of acceleration data which varied from 40 to 200 Hz. Moreover, accelerometer values differed in the provided value range and units. Visual mining of the data sets also showed some errors and non-uniformities from data sets, these are listed in Table 9.2. For instance, there are cases where sensor orientation of one study subject is not the same than for others.

9.2.2 Unifying Data Sets

To combine and provide an easy access to the selected ten data sets, it was necessary in the first place to unify these data sets. The data sets presented in Table 9.1 comes

Table 9.3 Data set includes accelerometer data from 17 activities. However, some of these are overlapping

Activity ID	Activity	Amount of data (%)
1	Standing	15.6
2	Sitting	13.1
3	Lying	8.0
4	Idling (= sitting + standing)	0.4
5	Walking	19.9
6	Walking (inc. walking at stairs)	0.2
7	Walking stairs up	10.3
8	Walking stairs down	8.9
9	Walking at stairs (inc. up and down)	0.2
10	Running (inc. jogging)	10.4
11	Biking	4.8
12	Jumping	1.8
13	Sitting in car	1.9
14	Elevator up	1.1
15	Elevator down	0.9
16	Falling	0.7
99	Null	1.9

in multiple file formats. Moreover, the underlying folder structure was different in different data sets. In some cases, the whole data set is in one single file but often data are divided into multiple folders and files based on study subject ID, body position ID or activity labels. In fact, one of the main benefits of OpenHAR is that it provides code to load these without taking care of file formats, and the resulting data set has only one format.

OpenHAR also unifies activity labels. Currently labels can be numeral or strings, and a number or string can have different meaning in different data sets. OpenHAR provides all the labels in numerical format and these labels have only one meaning. Activity labels used in OpenHAR are presented in Table 9.3. However, some of these activities are overlapping, which needs to be noted when OpenHAR data are used. For instance, Siirtola et al. (2012) contained activity *idling*, which is a combination of sitting and standing, while in other data sets *sitting* and *standing* were considered as two separate activities. Similarly some data sets consider *walking*, *walking upstairs* and *walking downstairs* as separate activities and in some data sets those all three are considered as one activity called *walking*. The same goes to *elevator up* and *elevator down* activities, in some cases they are combined as *elevator*-activity (direction not defined). Altogether, OpenHAR consists of 17 labels. Table 9.3 shows that most common activities are walking (19.9% of all the data), standing (15.6%) and sitting (13.1%).

Table 9.4 OpenHAR includes data from 14 body positions. However, some of these are overlapping

Position ID	Position	Amount of data (%)
1	Hip (inc. belt and waist)	22.5
2	Trouser's pocket, left (fixed orientation)	1.3
3	Trouser's pocket, right (fixed orientation)	1.3
4	Trouser's pocket, any (inc. thigh)	22.2
5	Chest	7.0
6	Wrist, any (inc. forearm)	10.1
7	Upper arm	7.7
8	Head	6.4
9	Shin (inc. leg)	13.3
10	Ankle	0.5
11	Trouser's pocket, left (free orientation)	0.5
12	Trouser's pocket, right (free orientation)	0.5
13	Foot, left	3.4
14	Foot, right	3.4

OpenHAR unifies sensor position IDs as well, the used position ID's are listed in Table 9.4. Sensor position is important to know as the measured sensor values are greatly dependent on the body position of the sensor. This means that if the model is trained using data from one position and tested using data from another position, the model does not work as wanted (Widhalm et al. 2018). Therefore, each observation of OpenHAR has a cell defining from which body position the value has been measured. It is worth noting that some of these positions are overlapping. For instance, in some studies sensor position was defined as trouser's pocket, meaning that it can either left or right, while in some cases position was explicitly defined as left of right pocket. In addition, in some cases study subjects were allowed to decide the orientation of the sensor while in some studies orientation was fixed. Moreover, some of the body positions were combined as they are so similar, for instance hip, waist and belt positions were combined as one. Most of the data is from hip (22.5%) and trouser's pocket (22.2%, including also data from thigh).

One feature of OpenHAR is that it unifies sampling rates depending on which data sets are combined. Unifying is based on down-sampling by finding the greatest common divider of the sampling rates of the selected data sets. Table 9.1 shown that sampling rates of the original data sets varied from 40 to 200 Hz. Therefore, if all ten data sets are combined, the sampling rate of the resulting data set is 10Hz, which has been shown to be enough to reliably recognize activities (Siirtola et al. 2011; Kose et al. 2012; Suto 2018).

Data sets were collected using multiple different sensors, and therefore, data sets had different value range and units. OpenHAR converts all the units as m/s^2 to enable the joint usage of the data sets. In addition, the range of measurements is unified.

All the data sets were visually mined before they were combined. Visual mining of the data sets showed some errors and non-uniformities from data sets, see Table 9.2. For instance, there are cases where sensor orientation of one study subject is not the same than for others. These were corrected by changing the common coordinate system to all data files within a data set. However, it should be noted that the orientation of a sensor is the same only within original data sets but when two original data sets are compared, the orientation of a sensor can differ. Nevertheless, this orientation issue can be solved for example using features extracted from magnitude signal, which is automatically calculated by OpenHAR using formula $\sqrt(\mathbf{x}^2 + \mathbf{y}^2 + \mathbf{z}^2)$, where **x**, **y** and **z** are the raw acceleration measurements from 3D accelerometer. Moreover, only a few original data set has timestamps, these were added to all data sets based on the sampling rate and by considering that it remains constant.

As a conclusion, OpenHAR contains over 65 million labeled data samples. This is equivalent to over 280 h of data from 3D accelerometers. This includes data from 211 study subjects, see Fig. 9.1. While the amount of data from each study subject varies a lot between data set (minimum 2 min, maximum 710 min), on average there is 80 min of data from each subject.

9.3 Using OpenHAR

OpenHAR is available from our research group's web page.[1] After downloading the toolbox, the next step is to download and unpack all ten original data sets (see ReadMe.txt—file to find instructions from where to download them and where to unpack them). After this, everything is set. To get the whole OpenHAR experience, Matlab is required. However, users without Matlab can also use OpenHAR. Unified data from all the ten data sets is available as a single .txt-file. Therefore, the sampling rate of this data set is 10 Hz. However, users without Matlab cannot have the whole OpenHAR experience as they have access to the data, but cannot use the commands presented in this section. This data file can also be downloaded from our research group's web page.

Using OpenHAR is easy, only one command is needed to download all the data to *data*-file. (`[data, sampling_rate] = getOpenHAR()`). In addition, this command returns the sampling rate of data. In this case, all the data are downloaded and so the sampling rate would be 10Hz. In fact, this is the data that is provided as a single .txt-file for users without Matlab.

User can also load only wanted parts of the whole data set by three name-value-pairs arguments (`'datasets'`, `'activities'`, and `'positions'`). User can specify several or only one name and value pair argument in any order.

If the purpose is not to use data from all ten original data sets, wanted data sets can be specified as the comma-separated pair consisting of `'datasets'` and a vector containing the IDs of the wanted data sets, see Table 9.1 for dataset IDs. For example,

[1]OpenHAR is available at: https://www.oulu.fi/bisg/datasets.

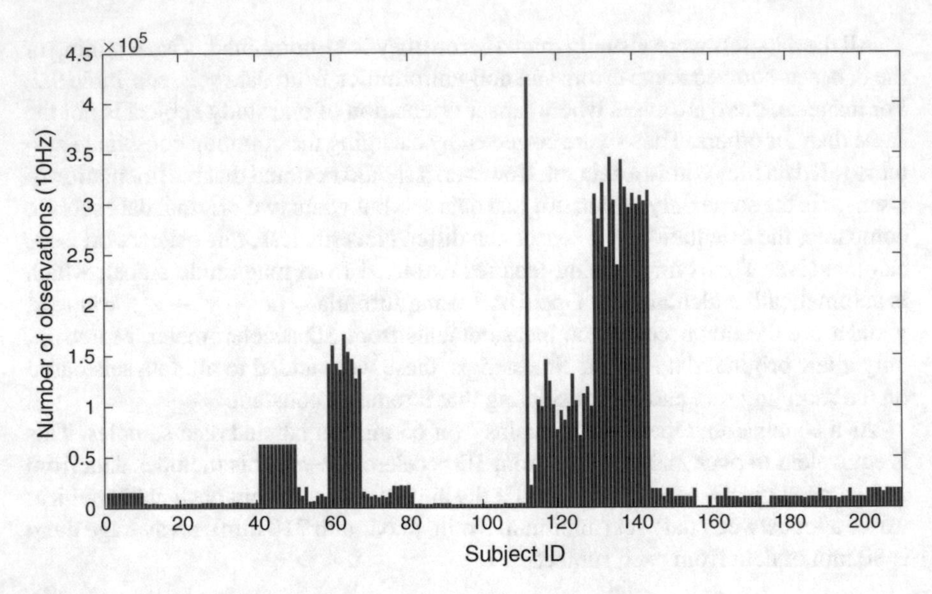

Fig. 9.1 OpenHAR contains data from 211 study subjects

the command `[data, sampling_rate] = getOpenHAR('datasets', [1 3 5])` returns only measurements from original data sets corresponding to data set IDs 1 (Banos et al. 2014), 3 (Shoaib et al. 2014), and 5 (Stisen et al. 2015). If all the selected data sets do not have the same sampling frequency, OpenHAR unifies them and return the sampling rate of the combined data set.

Similarly, if only some activities or body positions are of interest, wanted activities and body positions can be specified as the comma-separated pair consisting of `'activities'` and `'positions'`, and a vector containing the IDs of the wanted activities or body positions. IDs for these are listed in Tables 9.3 and 9.4. This means that user can select only data from some activities or body positions, or only data from some activities from selected body positions.

Each case, the code returns *data*-file. This file has nine columns of data in the following order: *data set ID, position ID, user ID, activity ID, timestamp, x-axis acceleration, y-axis acceleration, z-axis acceleration* and *magnitude acceleration*. File does not have a header.

When OpenHAR is used, we encourage users to cite to this chapter, and in addition, to cite to the publications were the original data sets were introduced. BibTeX-information for OpenHAR and original articles are provided in the ReadMe.txt—file of OpenHAR Matlab-package.

9.4 Experiments

This section shows how OpenHAR can be used to build reliable models for human activity recognition. Models are tested using one data set of OpenHAR and trained using the rest nine data sets. Experiments are done using three different classifiers: linear discriminant analysis (LDA), quadratic discriminant analysis (QDA) and classification and regression tree (CART). LDA is used to find a linear combination of features that separate the classes in an optimal way. The resulting combination and the the hyperplane separating the classes can then be employed as a linear classifier. QDA is a similar method but uses quadratic hyperplanes to separate classes (Hand et al. 2001). CART is a standard decision tree, which uses certain criterion to partition the space spanned by the input variables to maximize the score of class purity (Breiman 2017). Moreover, as sensor orientation within the ten data sets of OpenHAR can differ, it was decided to use only features extracted from the magnitude acceleration signal in the machine learning process. Altogether, 17 features were extracted from the magnitude acceleration signal. These included standard deviation, mean, maximum, minimum, different percentiles, and frequency domain features.

In the experiment, data from Shoaib et al. (data set ID 3) was use for testing, and the rest of the OpenHAR data for training. Therefore, the original testing data includes eight activities: walking, running, sitting, standing, jogging, biking, walking upstairs and walking downstairs. However, for the experiment walking, walking upstairs, and walking downstairs were combined as one walking-activity, and sitting and standing were combined as idling. The same was done for the training data, and this way it was possible to use the rest nine data sets for training. Moreover, in OpenHAR jogging is labeled as running. Thus, the testing data sets contained four activities: walking, running, idling, and biking. In addition, Shoaib et al. contains data from left and right trouser's pocket, wrist, upper arm, and belt positions. Therefore, body position IDs 1, 2, 3, 4, 5, 6, 7, 8, 11, and 12 were selected for the training data set.

Training was performed so that in the first place only one out of nine training data sets is used for training, and this model is tested using our test data set. Then incrementally more data sets, one at the time, are added to the training data set. Moreover, as a comparison, Shoaib et al. is classified using leave-one-subject-out-method. This means that in turn one persons data is used for testing and data from other persons for training.

Shoaib et al. contains data from 10 persons, and the recognition results were calculated separately to each person. The average accuracy from these was calculated and the results are presented in Figs. 9.2, 9.3, and 9.4. In addition, variance from these accuracies was calculated but variance remained practically the same no matter which training data was used. The results presented in the figures show that the best classification results are obtained using more than one OpenHAR data set. Therefore, it is beneficial to incrementally add more and more data sets to obtain the highest possible recognition accuracy. Moreover, it can be noted that with using LDA and QDA classifiers with OpenHAR data it is possible to obtain as high or even higher recognition rates in a previously unseen test data than by dividing this unseen data for test and training data. This means that instead of using time for collecting and

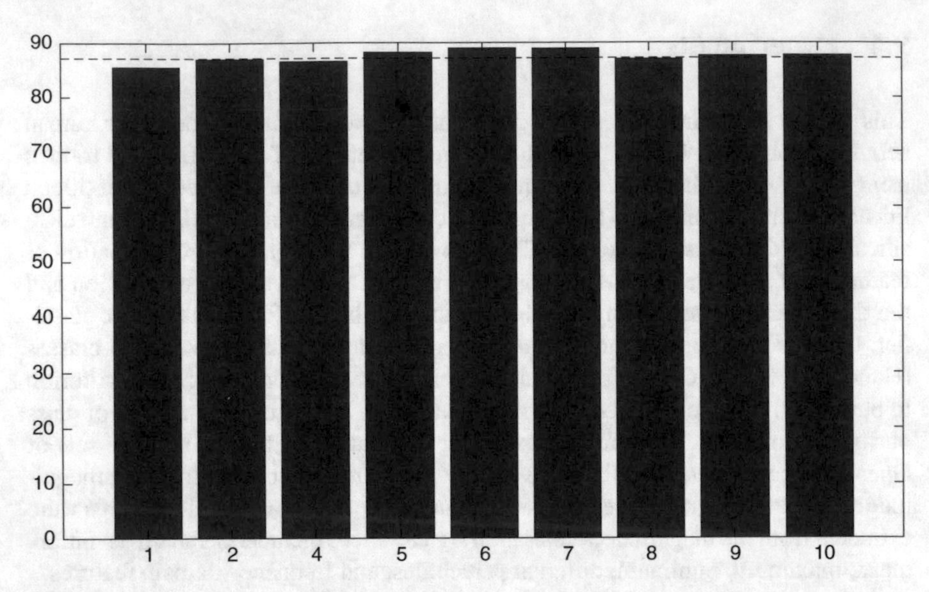

Fig. 9.2 Classification accuracies for Shoaib et al. data set using LDA classifier, when new OpenHAR data sets are incrementally added to training data. Bar number shows the data set ID of the added data set. Horizontal line shows the mean accuracy obtained by classifying Shoaib et al. using leave-one-subject-out-method

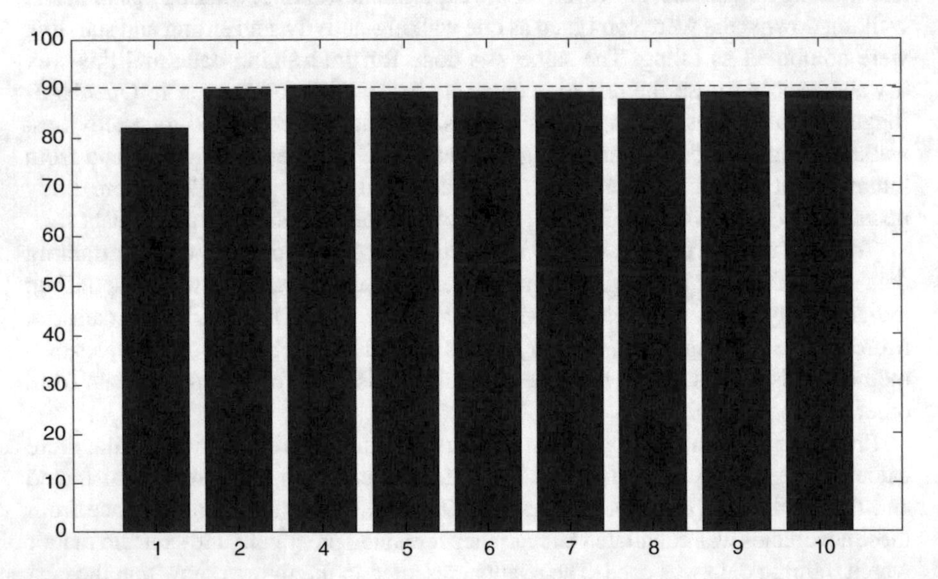

Fig. 9.3 Classification accuracies for Shoaib et al. data set using QDA classifier, when new OpenHAR data sets are incrementally added to training data. Bar number shows the data set ID of the added data set. Horizontal line shows the mean accuracy obtained by classifying Shoaib et al. using leave-one-subject-out-method

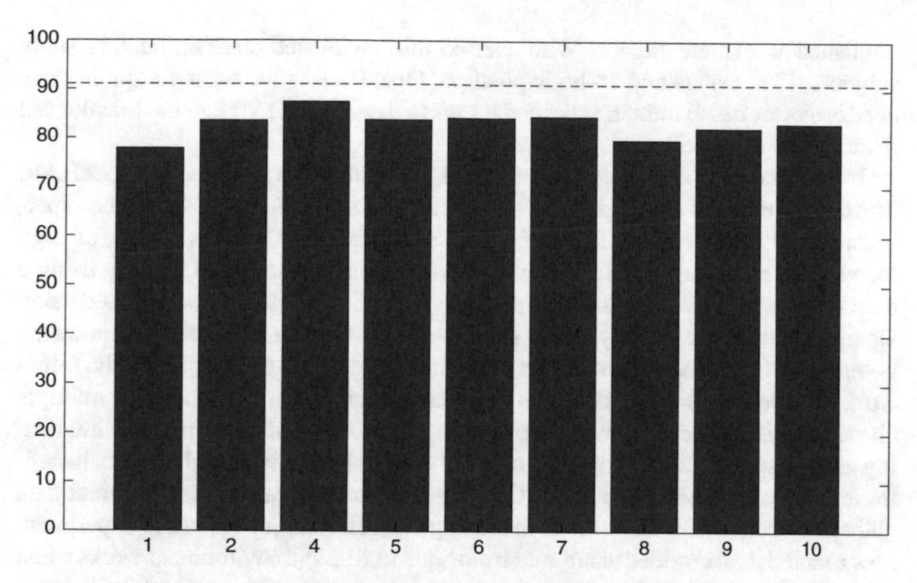

Fig. 9.4 Classification accuracies for Shoaib et al. data set using CART classifier, when new OpenHAR data sets are incrementally added to training data. Bar number shows the data set ID of the added data set. Horizontal line shows the mean accuracy obtained by classifying Shoaib et al. using leave-one-subject-out-method

labeling accelerometer data set, one can save time and just download OpenHAR and use the data provided by it. On the other hand, it should be noted that with CART the results obtained using OpenHAR data were slightly lower than with other two classifiers. However, also when using CART, in the best case (using data set IDs 1, 2, and 4 as training data) the difference in accuracies was not big when comparing results obtained, 91.0% versus 87.4%, respectively.

The experiment showed in this section was limited, but still it shows the potential of OpenHAR. However, to show the full potential of OpenHAR, more experiment should be done with multiple classifiers, including deep learning methods.

9.5 Discussion and Conclusion

OpenHAR is a free Matlab toolbox combining ten publicly available human activity data sets. The extra value provided by OpenHAR is that it provides easy access to these ten data sets. In fact, OpenHAR provides all the data sets in the same format. In addition, units, measurement range and labels are unified, as well as, body position IDs. Moreover, data sets with different sampling rates are unified using downsampling. What is more, data sets have been visually inspected to find visible errors, such as sensors in wrong orientation. OpenHAR improves re-usability of data sets by fixing these errors. Matlab code is provided to enable an easy way to select only desired parts of these data sets. In fact, in this chapter OpenHAR is introduced also to users without Matlab. They now have an access to OpenHAR data as it is published as a single .txt-file. With over 65 million labeled observation, 211 study subjects, 17 activities and 14 body position, OpenHAR is the most comprehensive accelerometer based human activity data set to date. OpenHAR can be downloaded from https://www.oulu.fi/bisg/datasets.

In this chapter, a small experiment was done to show the potential on OpenHAR. The experiment was done using three classifiers: LDA, QDA, and CART. The experiment showed that using LDA and QDA classifiers and OpenHAR data, as high recognition rates can be achieved in a previously unseen test data than by using a data set specially collected for this purpose. With CART the results obtained using OpenHAR data were slightly lower. The experiment was limited, but still it shows the potential of OpenHAR. However, to show the full potential of OpenHAR, the future work includes more comprehensive experiments and experimenting with multiple classifiers. Experimenting with deep learning methods would be especially interesting as they are data hungry, and the size of OpenHAR data would, therefore, benefit these types of models. In addition, OpenHAR contains data from ten different data gathering protocols, which means that using OpenHAR is it possible to experiment how a model that is trained using data from one location and environment works when it is tested in other location. Thus, future work includes testing methods of transfer learning. Moreover, new data sets can be added to OpenHAR in the future when suitable data set are published. In addition, currently OpenHAR focuses on daily activities, however, it could be extended to include other types of activities as well.

Acknowledgements The authors would like to thank Infotech Oulu for funding this work. In addition, the authors would like to thank the authors of Anguita et al. (2013), Banos et al. (2014), Chereshnev and Kertész-Farkas (2017), Micucci et al. (2017), Shoaib et al. (2014), Siirtola et al. (2012), Stisen et al. (2015), Sztyler and Stuckenschmidt (2016), Vavoulas et al. (2016), Zhang and Sawchuk (2012) for collecting and publishing the original data sets.

References

Anguita D, Ghio A, Oneto L, Parra Perez X, Reyes Ortiz JL (2013) A public domain dataset for human activity recognition using smartphones. In: Proceedings of the 21th international European symposium on artificial neural networks, computational intelligence and machine learning, pp 437–442

Banos O, Garcia R, Holgado-Terriza JA, Damas M, Pomares H, Rojas I, Saez A, Villalonga C (2014) mHealthDroid: a novel framework for agile development of mobile health applications. In: Pecchia L, Chen LL, Nugent C, Bravo J (eds) Ambient assisted living and daily activities. Springer International Publishing, Cham, pp 91–98

Bartlett J, Prabhu V, Whaley J (2017) Acctionnet: a dataset of human activity recognition using on-phone motion sensors. In: Proceedings of the 34th international conference on machine learning, Sydney, Australia

Breiman L (2017) Classification and regression trees. Routledge

Chereshnev R, Kertész-Farkas A (2017) HuGaDB: human gait database for activity recognition from wearable inertial sensor networks. In: International conference on analysis of images, social networks and texts. Springer, pp 131–141

Gjoreski H, Ciliberto M, Wang L, Morales FJO, Mekki S, Valentin S, Roggen D (2018) The university of Sussex-Huawei locomotion and transportation dataset for multimodal analytics with mobile devices. IEEE Access

Hand DJ, Mannila H, Smyth P (2001) Principles of data mining. MIT Press, Cambridge, MA, USA

Incel O, Kose M, Ersoy C (2013) A review and taxonomy of activity recognition on mobile phones. BioNanoScience 3(2):145–171

Kose M, Incel OD, Ersoy C (2012) Online human activity recognition on smart phones. In: Workshop on mobile sensing: from smartphones and wearables to big data, vol 16, pp 11–15

Lockhart JW, Pulickal T, Weiss GM (2012) Applications of mobile activity recognition. In: 2012 ACM conference on ubiquitous computing, UbiComp '12. ACM, New York, NY, USA, pp 1054–1058

Micucci D, Mobilio M, Napoletano P (2017) UniMiB SHAR: a dataset for human activity recognition using acceleration data from smartphones. Appl Sci 7(10):1101

Reiss A, Stricker D (2012) Introducing, a new benchmarked dataset for activity monitoring. In: 16th international symposium on wearable computers (ISWC). IEEE, pp 108–109

Shoaib M, Bosch S, Incel OD, Scholten H, Havinga P (2014) Fusion of smartphone motion sensors for physical activity recognition. Sensors 14(6):10146–10176

Siirtola P, Komulainen J, Kellokumpu V (2018a) Effect of context in swipe gesture-based continuous authentication on smartphones. In: 26th European symposium on artificial neural networks, computational intelligence and machine learning, ESANN 2018, pp 639–644, 25–27 Apr 2018

Siirtola P, Koskimäki H, Röning J (2018b) Personalizing human activity recognition models using incremental learning. In: 26th European symposium on artificial neural networks, computational intelligence and machine learning, ESANN 2018, pp 627–632, 25–27 Apr 2018

Siirtola P, Koskimäki H, Röning J (2018c) Experiences with publicly open human activity data sets—studying the generalizability of the recognition models. In: Proceedings of the 7th international conference on pattern recognition applications and methods. SCITEPRESS, pp 291–299

Siirtola P, Koskimäki H, Röning J (2018d) OpenHAR: a matlab toolbox for easy access to publicly open human activity data sets. In: Proceedings of the 2018 ACM international joint conference and 2018 international symposium on pervasive and ubiquitous computing and wearable computers. ACM, pp 1396–1403

Siirtola P, Laurinen P, Röning J, Kinnunen H (2011) Efficient, accelerometer-based swimming exercise tracking. In: IEEE Symposium on computational intelligence and data mining (CIDM). IEEE, pp 156–161

Siirtola P, Röning J (2012) Recognizing human activities user-independently on smartphones based on accelerometer data. Int J Interact Multimed Artif Intell 1(5):38–45 June

Stisen A, Blunck H, Bhattacharya S, Prentow TS, Kjærgaard MB, Dey A, Sonne T, Jensen MM (2015) Smart devices are different: Assessing and mitigatingmobile sensing heterogeneities for activity recognition. In: Proceedings of the 13th ACM conference on embedded networked sensor systems. ACM, pp 127–140

Suto J, Oniga S, Lung C, Orha I (2018) Comparison of offline and real-time human activity recognition results using machine learning techniques. In: Neural computing and applications, pp 1–14

Sztyler T, Stuckenschmidt H (2016) On-body localization of wearable devices: an investigation of position-aware activity recognition. In: 2016 IEEE international conference on pervasive computing and communications (PerCom). IEEE, pp 1–9

Vavoulas G, Chatzaki C, Malliotakis T, Pediaditis M, Tsiknakis M (2016) The mobiact dataset: recognition of activities of daily living using smartphones. In: ICT4AgeingWell, pp 143–151

Widhalm P, Leodolter M, Brändle N (2018) Top in the lab, flop in the field? Evaluation of a sensor-based travel activity classifier with the SHL dataset. In: Proceedings of the 2018 ACM international joint conference and 2018 international symposium on pervasive and ubiquitous computing and wearable computers, UbiComp '18. ACM, New York, NY, USA, pp 1479–1487

Zhang M, Sawchuk AA (2012) USC-HAD: a daily activity dataset for ubiquitous activity recognition using wearable sensors. In: ACM international conference on ubiquitous computing (Ubicomp) workshop on situation, activity and goal awareness (SAGAware), Pittsburgh, Pennsylvania, USA, Sept 2012

Chapter 10
MEASURed: Evaluating Sensor-Based Activity Recognition Scenarios by Simulating Accelerometer Measures from Motion Capture

Paula Lago, Shingo Takeda, Tsuyoshi Okita and Sozo Inoue

Abstract Human Activity Recognition from accelerometer sensors is key to enable applications such as fitness tracking or health status monitoring at home. However, evaluating the performance of activity recognition systems in real-life deployments is challenging to the multiple differences in sensor number, placement and orientation that may arise in real settings. Considering such differences requires a large amount of labeled data. To overcome the challenges and costs associated to the collection of a wide range of heterogeneous data, we propose a simulator, called MEASURed, which uses motion capture to simulate accelerometer data on different settings. Then, using the simulated data to estimate the performance of activity recognition models under different scenarios. In this chapter, we describe MEASURed and evaluate its performance in estimating the accuracy of activity recognition models. Our results show that MEASURed can estimate the average accuracy of an activity recognition model using real accelerometer magnitude data. By using motion capture to simulate accelerometer data, the sensor research community can profit from visual datasets that have been collected by other communities to evaluate performance of activity recognition in a wide range of activities. MEASURed can be used to evaluate activity recognition classifiers in settings with different number, placement, and sampling rate of accelerometer sensors. The evaluation on a broad spectrum of scenarios gives a more general view of models and their limitations.

P. Lago (✉) · S. Takeda · T. Okita · S. Inoue
Kyushu Institute of Technology, Kitakyushu-shi, Fukuoka, Japan
e-mail: paula@mns.kyutech.ac.jp

S. Takeda
e-mail: takeda@sozolab.jp

T. Okita
e-mail: tsuyoshi@mns.kyutech.ac.jp

S. Inoue
e-mail: sozo@mns.kyutech.ac.jp

N. Kawaguchi et al. (eds.), *Human Activity Sensing*,
Springer Series in Adaptive Environments,
https://doi.org/10.1007/978-3-030-13001-5_10

10.1 Introduction

Human Activity Recognition (HAR) from inertial sensors enables a wide range of applications in different domains such as healthcare, surveillance and smart homes (Lara and Labrador 2013). HAR is usually based in supervised machine learning, therefore data collection, data preprocessing, model training and evaluation are important steps in this area (Bulling et al. 2014). While research work often report high accuracy results, the same results are often not achieved in real-world deployment, because the impact of technology differences is not considered in the laboratory setting (Amft 2005). Differences in sensing devices and technology such as sampling rate, sensing capabilities (value range), position and orientation can have an enormous impact on the final model accuracy (Wüstenberg et al. 2013). Although it is desired to report how such differences can impact activity recognition models (Amft 2005), the cost of collecting data makes comprehensive evaluation of a system a big challenge for the community (Bulling et al. 2014).

Previously, public activity datasets using accelerometers (Jeffrey et al. 2011; Daniela et al. 2017) have been released by the HAR community. However, such datasets are measured with few or only one sensor placements and sensor specifications, hampering system assessment under various scenarios. Multi-accelerometer datasets (Gjoreski et al. 2017; Stisen et al. 2015; Roggen et al. 2010) are becoming available but their size and variability in sensing characteristics may not be enough to make general conclusions about several changing dimensions.

To overcome these challenges, we propose to use motion capture data to simulate accelerometer data (Sect. 10.3) and use simulated data to compare different scenarios for activity recognition (Sect. 10.4). We notice several motion capture data sets are available providing large collections of actions and multiple marker placements (Sect. 10.5). Moreover, motion capture datasets typically feature high sampling rates, much higher than those of accelerometer sensors. We observe that it is possible to derive acceleration from motion capture since it provides position data. By using motion capture to derive acceleration, it is possible to simulate different sensor placements, sampling rates and rotations.

This chapter is an extended version of Takeda et al. (2018) where we proposed MEASURed (MultisEnsor Activity recognition SimUlatoR), a simulator to benefit from open motion capture data for such purposes (Sect. 10.2). In this chapter, we describe in detail the challenges of simulating accelerometer data and provide an extensive evaluation of the method followed to do this. The main contribution of this work is:

- MEASURed allows comparing the performance of activity recognition under different sensor conditions without having to collect new datasets.

Next, we provide an overview of the proposed system.

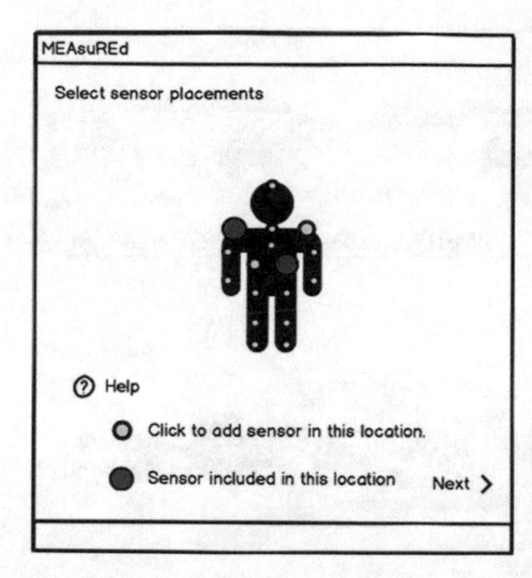

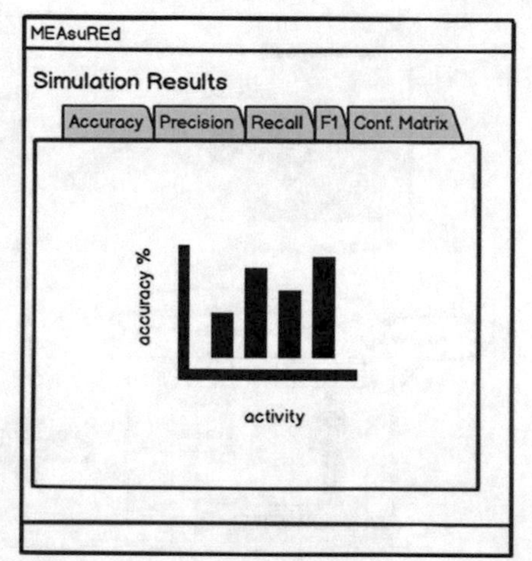

Fig. 10.1 User selects the number and placement of sensors, as well as their sampling rate, and MEASURed outputs the estimated performance of an Activity Recognition model using such settings

10.2 MEASUred Overview

We propose to use optical motion capture datasets to simulate the performance of an activity recognition system using accelerometers. A motion capture system records the movements of a person. Optical motion capture can track the position of a high number of marker placements at high sampling rates. Thanks to this information, it is possible to derive acceleration data for each marker.

Each marker can be used to simulate a sensor at that position. Therefore, MEASURed allows the user to select a sampling rate and the desired number and placement of accelerometer sensors. Using the simulated accelerometer data, MEASURed then provides an estimation of the performance of a classifier using the selected settings (Fig. 10.1).

MEASURed follows two steps (Fig. 10.2): (1) simulating accelerometer data from motion capture for the selected sensor positions and sampling rate and (2) simulating activity recognition performance for the chosen setting.

In what follows, we detail the two main steps of MEASURed.

10.3 Simulating Accelerometer Data from Motion Capture Data

The first step in MEASURed is to simulate the accelerometer data. For each chosen sensor position, measurements at the specified sampling rate are determined from the position data obtained from the motion capture. Linear acceleration is the second

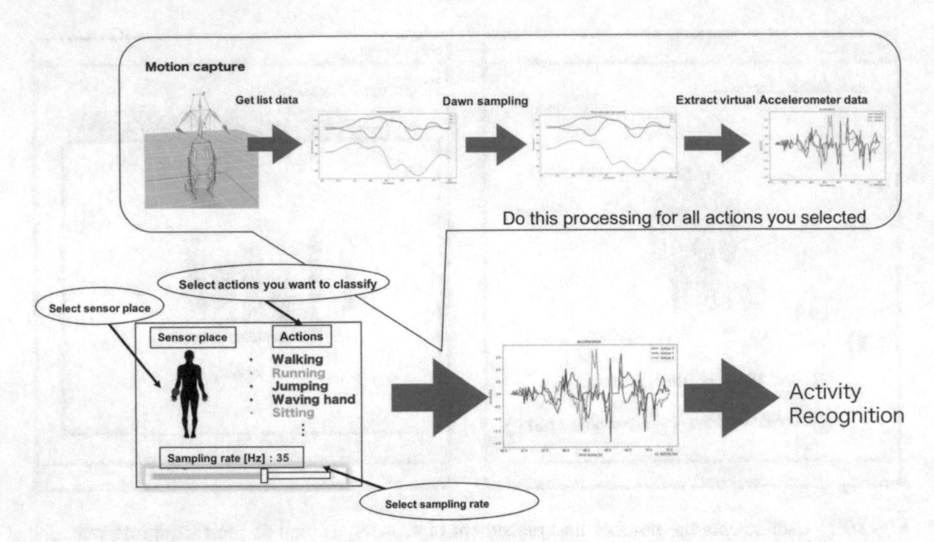

Fig. 10.2 MEASURed algorithm follows two steps: simulating acceleration data and simulating activity recognition model performance

derivative of the position data, but the nature of accelerometer sensors raise issues of orientation and axis alignment. In this section, we first explain the factors involved in the simulation of accelerometer data (Sect. 10.3.1). We then explain two methods to simulate the acceleration, by magnitude (Sect. 10.3.2) and by axis (Sect. 10.3.3).

10.3.1 Linear Acceleration and Proper Acceleration

We can easily obtain an acceleration value from motion capture data by calculating the second difference of the position data. However, this value differs from the acceleration measured by an accelerometer sensor due to their different reference frame. Accelerometers measure the proper acceleration of the sensor, that is the acceleration with respect to its instantaneous rest reference frame. In contrast, motion capture systems measure position with respect to a global reference frame (Fig. 10.3). Thus, the acceleration obtained by differentiation of the position is a coordinate acceleration, with respect to the global reference frame.

Because the accelerometer sensor measures the acceleration with respect to its own rest reference frame, its measurements are always influenced by the force of gravity. A motionless accelerometer measures 1g positive acceleration. But, the acceleration derived from motion capture is not subject to such force. To be able to compare both accelerations, we need to: (1) remove the effect of gravity from the accelerometer measures and (2) transform their reference frames to be equal.

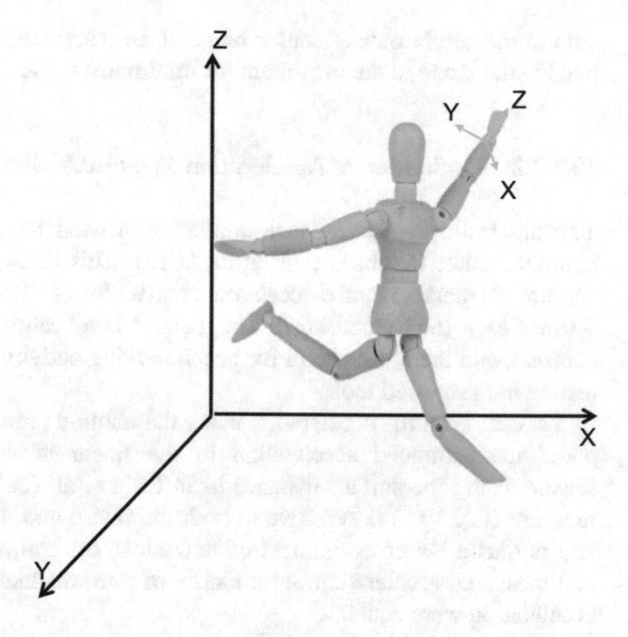

Fig. 10.3 Motion capture measures position with respect to a global reference frame. Accelerometer sensors measure acceleration with respect to the sensor's rest-frame

In the following we consider two cases. First, we simulate only accelerometer magnitude. In such case we need to only consider the influence of the force of gravity since the reference frame rotations will not affect the final magnitude. In the second case, we simulate each axis acceleration and provide some considerations on how to deal with frame rotations.

10.3.2 Estimating Acceleration Magnitude

When estimating acceleration magnitude we only need to consider the effect of gravity into the accelerometer sensor measurements. We simulate the linear acceleration and remove the effect of gravity from the accelerometer sensor measurements.

10.3.2.1 Estimating Magnitude of Linear Acceleration Using Motion Capture Data

To estimate acceleration magnitude, we first obtain the magnitude of each position point. Then, we simulate a linear acceleration in the desired sampling rate. For this, we use a decimation process involving a low-pass filter and down-sampling by a factor of $M = m_r/d_r$ where m_r is the motion capture sampling rate and d_r is the desired sampling rate. After decimation, we obtain the acceleration data by using a Savitzky-Golay filter as in Ruiz et al. (2018). Finally, we need to adjust the derived data to the accelerometer sensor range of measurement. For this, we set all values outside the range to the maximum (or minimum) value.

10.3.2.2 Evaluation of Acceleration Magnitude Simulation

To evaluate the precision of such simulation we used the public Berkeley Multimodal Human Action Database (Ofli et al. 2013). This dataset contains optical motion capture (43 markers) and 6-accelerometer data for 12 subjects performing 11 actions, 5 times each (657 recordings). This dataset is a "carefully designed experimental dataset", and therefore useful for benchmarking activity recognition proposals and testing our proposed tool.

We estimated the acceleration using the method previously described. We compared the estimated acceleration to the linear acceleration measured by the sensor[1] using the surface distance as in Ruiz et al. (2018). The surface similarity measure (Eq. 10.1) is sensitive to both the phase and the magnitude of the signal and is thus a better assessment of how close the estimated acceleration is to the real measured acceleration. A similarity of 1 means that the estimated and the real acceleration were equal.

$$surf - sim(x, y) = 1 - \frac{||x - y||}{||x|| + ||y||} \tag{10.1}$$

The average similarity of all estimated accelerations for each sensor position is shown in Fig. 10.4 with their standard deviation. From the data in Fig. 10.4 we can conclude that the estimated accelerations were highly similar to the accelerometer measured acceleration. As an example, we show some of the signals with the highest similarity and some of the signals with the lowest similarity in Fig. 10.5.

10.3.3 Estimating Acceleration by Axis

To simulate the acceleration by axis we must follow the same steps as the simulation of magnitude. That is, a decimation process followed by a derivative filter. This gives the linear acceleration in each axis on the global coordinate frame determined by the motion capture system. To obtain the acceleration in the reference frame of the accelerometer, we need to consider its orientation. Sensor orientation affects the measurements of the inertial sensor because all axes rotate as it moves.

Part of the orientation of the accelerometer is given by the rotation of the rigid body where it is placed. As an example, if it is in the wrist its rotation will be given by the arm rotations (Fig. 10.6). Such rotation angles are determined by the motion

---

[1] We obtain linear acceleration using a high pass filter as described in https://developer.android.com/guide/topics/sensors/sensors_motion#java and apply a median filter to remove noise.

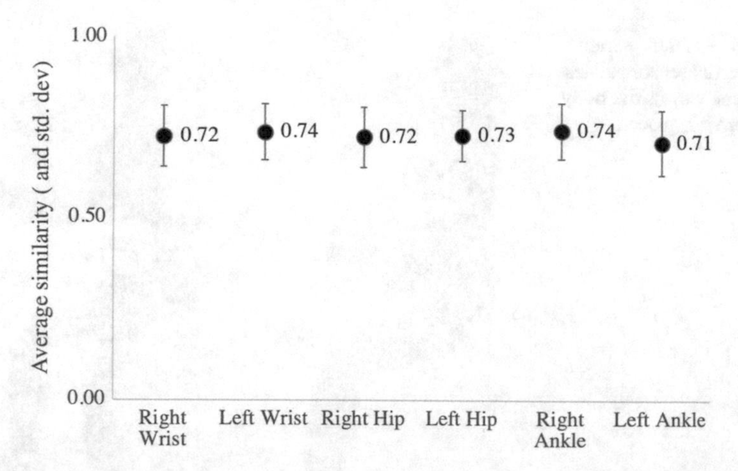

Fig. 10.4 Average similarity obtained for all estimated accelerations in the MHAD Dataset and their standard deviation

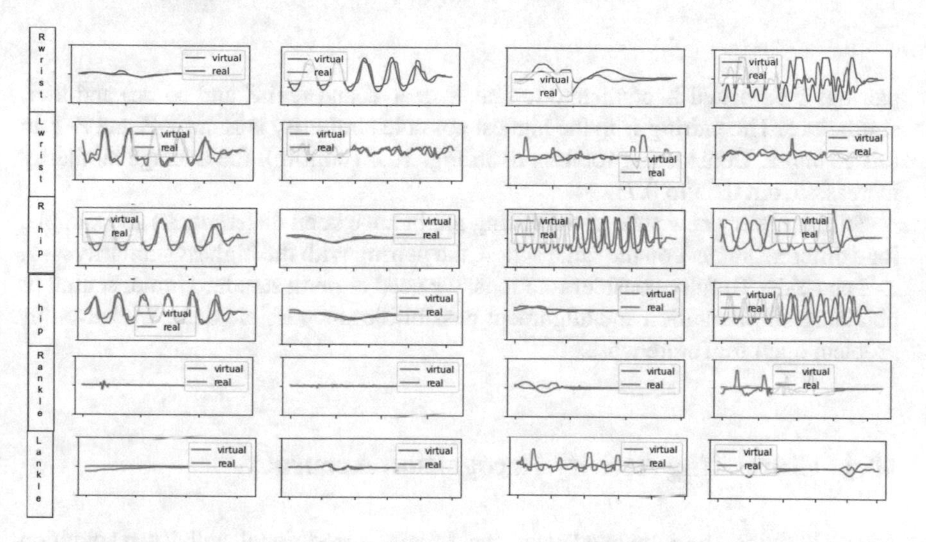

Fig. 10.5 Real and estimated acceleration magnitude for different sensor positions. The two left-most columns show examples of high similarity and the right-most columns

capture system using kinematics equations and given in the We used the bvh file of the dataset to obtain it.

However, we have no knowledge about the original placement of the axis and their positive and negative orientation. In the image of Fig. 10.6, any of the axes can be the X axis for example. This causes *axis alignment* differences. A different setting would give different results as the X axis in the coordinate frame does not correspond to the X axis in the body frame. To illustrate this, consider the linear accelerations obtain for the `throw` action shown in Fig. 10.7 (top). We show all possible axis

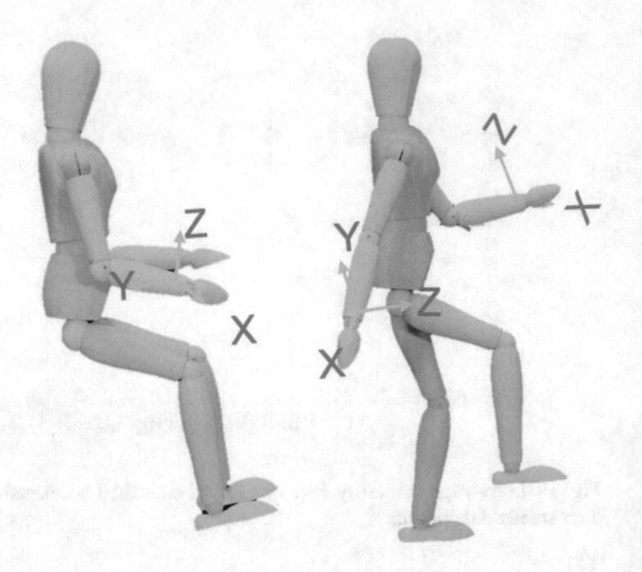

Fig. 10.6 Reference frame of the inertial sensor rotates in the same way as the body part where it is placed

pairings (i.e. virtual X compared to real X, real Y and real Z and so on) and their similarities. The pairing with the highest possible similarity is X-X, Y-Z and Z-Y in this example. Considering rotation as in Fig. 10.7 (bottom), the average similarity increases from 0.65 to 0.71.

Still, for every new subject the setting might have been different. As an example, for a different subject on the same action, the pairing with the highest similarity X-Y, Z-Z and Y-X. To fully consider rotations, we need either a standard initial setting or obtaining all orientation and alignment possibilities as possible data. We leave this problem open for future work.

10.4 Simulating Activity Recognition Accuracy

After obtaining the simulated data, we follow a traditional activity recognition pipeline to simulate the performance of a classifier. In MEASURed, the user can select the desired features to use and the classifier model. Currently, a set of possible features are given. These are statistical features that have been shown to provide good results in physical activity recognition: median, standard deviation, variance, skewness, kurtosis and time weighted variance.

The accuracy of a model using real accelerometer data is estimated using confidence intervals from the accuracy obtained using the simulated acceleration. The confidence intervals are calculated as $[acc - cv, acc + cv]$ where acc is the accuracy obtained by the estimated model and cv is the confidence interval value (Eq. 10.2).

$$cv = z * sqrt((acc * (1 - acc)/n) \tag{10.2}$$

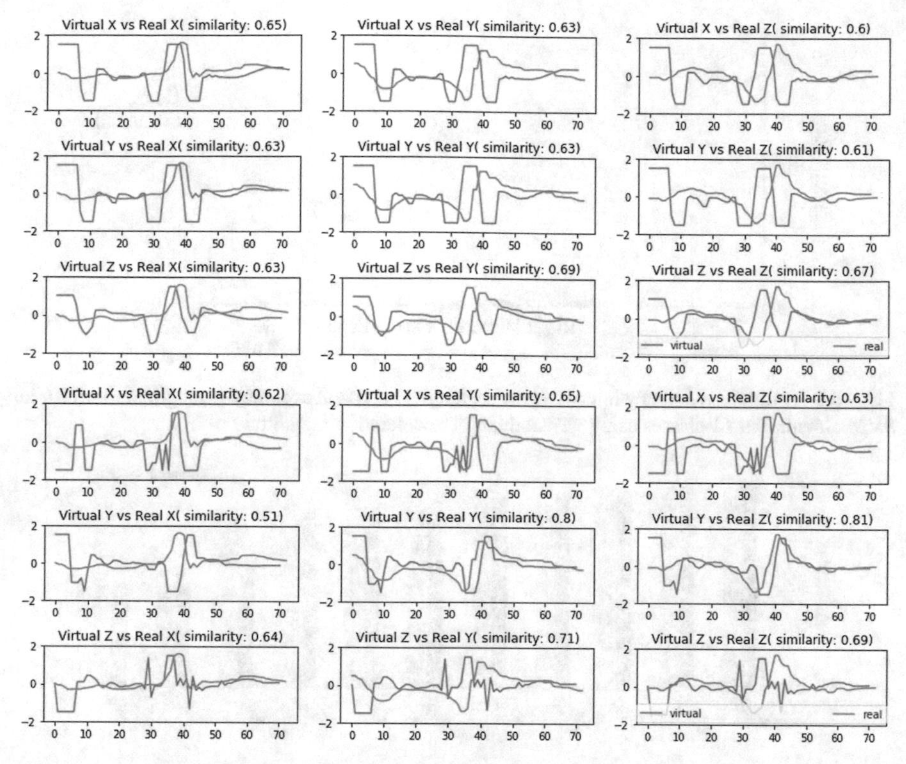

Fig. 10.7 Pairings of virtual and real accelerations when we consider no rotations (top) and rotations around Z, X, Y (bottom). The average similarity increased from 0.65 to 0.71. The axis pairing giving the highest similarity are X-X, Y-Z and Z-Y

where:

- z depends on the level of confidence of the interval. For a 95% confidence interval z = 1.96 and for a 98% confidence interval, z = 2.33.
- *acc* is the accuracy given by the model trained with simulated data
- n is the number of samples in the test data.

10.4.1 Evaluating Activity Recognition Simulation Using Acceleration Magnitude

We evaluated the activity recognition simulation using the accelerations obtained by simulation as explained in Sect. 10.3.2. We used all 7 statistical features and an SVM as the classifier. For all sensor positions, a random split of 75% of the data was used for training and 25% of the data (165 samples) was used for testing.

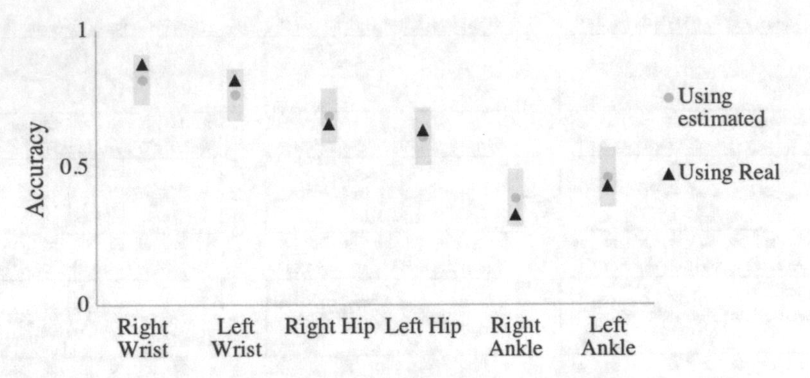

Fig. 10.8 Estimated accuracy interval (gray rectangles) and real accuracy (triangle marker) for an SVM classifier in 11 classes using virtual data for acceleration magnitude

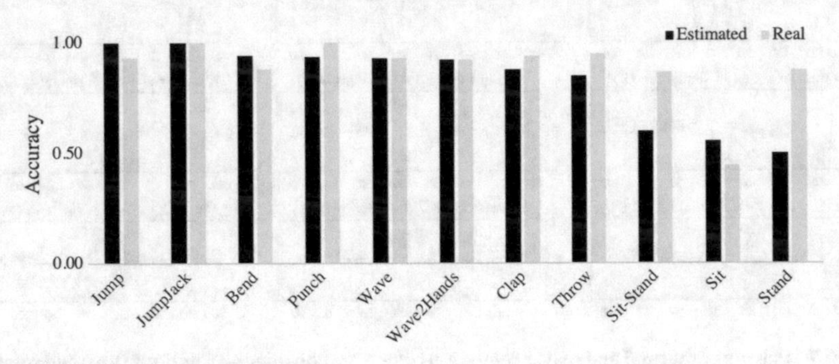

Fig. 10.9 Class accuracy for the right wrist sensor. The estimated accuracy for each class is not consistent with the real accuracy

In this evaluation, the real average accuracy obtained with real accelerometer data was always inside the estimated confidence interval (Fig. 10.8).

MEASURed provides a good estimation of the accuracy of a classifier. A limitation exists in estimating the accuracy of each activity class. Some activities' accuracy is overestimated while that of others is underestimated. This factor prevents from a clear overview of a classifier performance. As an example, we show the estimated and real class accuracy obtained for the right wrist sensor (Fig. 10.9).

10.4.2 Evaluating Activity Recognition Simulation Using Acceleration by Axis

We performed experiments to simulate activity recognition using virtual data by axis. For this, we used the rotations in the ZYX order and do not change any axis. We follow the same procedure as in the previous section, using the same features.

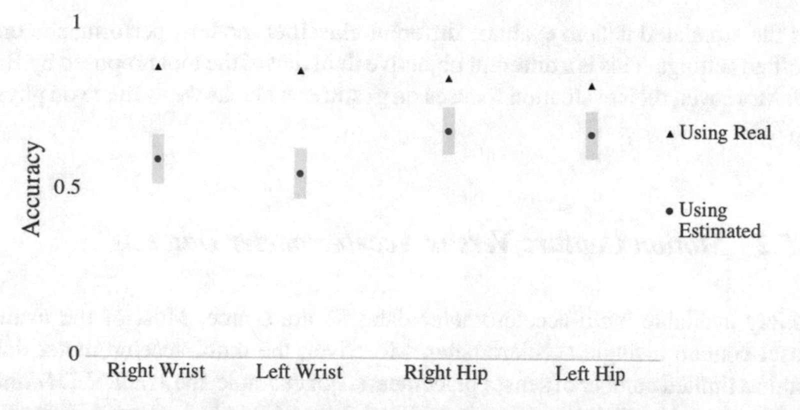

Fig. 10.10 Estimated accuracy interval (gray rectangles) and real accuracy (triangle marker) for an SVM classifier in 11 classes using virtual data for each axis

The results of the estimated and real accuracy are shown in Fig. 10.10. In this evaluation, the estimated accuracy was far from the real accuracy. Using axis alignment and trying different rotation orders, this might improve, but the work is left as future work.

10.5 Related Work

In this section, we first present related work in translating from one sensing modality to another sensing modality. Then, by presenting publicly available datasets for both modalities, we argue the advantage of using motion capture datasets in MEASURed, which are more diverse and numerous.

10.5.1 Translating Between Sensing Modalities

Accelerometer simulation tools that use motion capture for data derivation have been proposed (Asare et al. 2013; Young et al. 2011). They focus solely in achieving good accuracy of the simulated data. Yet they do not state what the objective of such translation is. The objective in MEASURed is to evaluate activity recognition in different settings. To provide accurate accelerometer measures, their simulation considers sensor orientation and sensor noise. MEASURed only considers linear acceleration but we use the simulated data for activity recognition.

In contrast, Banos et al. (2012) propose a method to translate between sensor modalities for activity recognition. Their goal is to use the same activity recognition model with different sensing modalities (inertial sensor and kinect). MEASURed uses the simulated data to evaluate different classifier models' performance on the specified settings. This is a different objective than that of the tool proposed by Banos et al. Moreover, their evaluation focuses on gestures whereas we evaluate on physical activities.

10.5.2 Motion Capture Versus Accelerometer Datasets

Publicly available multi-accelerometer datasets are scarce. Most of the available dataset contain a single accelerometer. Moreover, the multi-accelerometer dataset contain a limited number of sensor placements. For example, the *MHEALTH dataset*[2] has data about 10 subjects performing 12 activities with only 3 sensor positions. The *SHL Dataset* (Gjoreski et al. 2017) contains measurements from 4 sensor positions for 3 participants using 8 modes of locomotion. The *RealWorld HAR Dataset*[3] has data about 15 subjects performing 8 activities using 6 sensor positions. The *OPPORTUNITY Activity Recognition Dataset* (Roggen et al. 2010) has 19 accelerometers in different body positions. This is, to the best of our knowledge, the largest number of accelerometers used in any dataset. It has data of 4 users performing 5 high level activities each divided into actions (13 actions). The *Heterogeneity Human Activity Recognition Dataset (HHAR)* (Stisen et al. 2015) considers sensing heterogeneities of different devices and users. It contains the readings of accelerometer and gyroscopes using different device models. It records 9 users performing 6 activities. Although users carry different smartphone and smartwatch at the time, their positions are not given.

Motion Capture Datasets, in contrast, are highly available containing a wide range of activities and subjects even if they are not concerned with activity recognition research. The range of activities is more varied as well as the number of subjects. For example, the *HDM5 database*[4] contains data about 5 subjects performing 5 high level motions and annotations for 100 low-level motions in each of these classes. The *Open Motion Project*[5] has data about different activities, motions and performances for different subjects. The *library of human movements*[6] contains data of 30 subjects performing 4 activities in different styles and different emotional states. The *Dance Motion Capture Database*[7] contains motion capture data of Greek dances, contemporary dances and other performances with more than 200 videos. The *CMU Graphics Lab Motion Capture Database*[8] contains a wide selection of motion cap-

---

[2]https://archive.ics.uci.edu/ml/datasets/MHEALTH+Dataset.

[3]http://sensor.informatik.uni-mannheim.de/#dataset_realworld.

[4]http://resources.mpi-inf.mpg.de/HDM05/index.html.

[5]https://accad.osu.edu/research/motion-lab/system-data.

[6]http://paco.psy.gla.ac.uk/index.php?option=com_jdownloads&view=viewcategories&Itemid=62.

[7]http://dancedb.eu/main/performances.

[8]http://mocap.cs.cmu.edu.

ture data for 5 types of motions performed by 144 subjects most of them with several trials per motion. The *Vicon Physical Action Data Set*[9] contains data of 10 subjects performing 10 normal and 10 aggressive interactions.

10.5.3 Considering Sensor Orientations

Since sensor orientation has a strong impact in the accuracy of activity recognition methods (Chavarriaga et al. 2013), researchers have proposed techniques to avoid this impact. MEASUred uses the magnitude of the acceleration to obtain a better estimate of the accelerometer data since no information about the sensor orientation is available. This is similar to the method proposed by Kunze and Lukowicz (2014).

The impact of the changes in orientation comes from the influence of gravity in the measurements. To avoid this, Yang (2009) propose to estimate the direction of the gravity vector by averaging the acceleration vectors in the long term. Using this gravity vector and the magnitude of the orthogonal acceleration as features for activity recognition, they achieve an orientation invariant method.

When gyroscope and magnetometer sensors are available, the acceleration vectors can be transformed from inertial reference frame to an Earth-centered reference frame (Ustev et al. 2013; Yurtman et al. 2018). With this transformation, the direction of each axis becomes constant, as opposed to those of the inertial reference frame. As such, the features no longer depend on the sensor orientation.

10.6 Conclusions

In this chapter, we have presented MEASURed, as a tool to evaluate activity recognition systems under different scenarios using estimated accelerometer data. We use motion capture data to estimate the accelerometer data at each marker placement. By the use of motion capture data with a high number of markers and high sampling rates, MEASURed allows the user to specify the desired number of sensors as well as the location and sampling rate for each sensor. With this tool, the impact of these parameters is an activity recognition model can be fairly evaluated. The main advantage of MEASURed is that it profits from the available motion capture datasets which are far more heterogeneous and available than multi-accelerometer datasets.

Data collection for training activity recognition models is a costly task. In addition, given the variability of accelerometers used in smartphones, smartwatches and other wearable devices, assessing systems for real-life usage becomes very challenging. MEASURed can estimate the performance of activity recognition on different scenarios, and can be used to disclose limits of a given model.

[9]https://archive.ics.uci.edu/ml/datasets/Vicon+Physical+Action+Data+Set.

Simulated data from motion capture can also be used to estimate accelerometer data at different sensor orientations. The evaluation of the accuracy of estimated data at different orientations is difficult due to the lack of real data to compare with. Nonetheless, the current findings could indicate that simulated data can be used to evaluate the impact of sensor orientation also.

References

Amft O (2005) On the need for quality standards in activity recognition using ubiquitous sensors. In: UbiComp '13 Adjunct proceedings of the 2013 ACM conference on pervasive and ubiquitous computing adjunct publication, pp 62–79

Asare P, Dickerson RF et al (2013) BodySim: a multi-domain modeling and simulation framework for body sensor networks research and design. In: Proceedings of SenSys '13, pp 1–2

Banos O, Calatroni A, et al (2012) Kinect=imu? learning MIMO signal mappings to automatically translate activity recognition systems across sensor modalities. In: Proceedings of the 16th ISWCWashington, DC, USA, 2012. IEEE Computer Society, pp 92–99

Bulling A, Blanke U, Schiele B (2014) A tutorial on human activity recognition using body-worn inertial sensors. ACM Comput Surv 46(3):33:1–33:33

Chavarriaga R, Bayati H, Millán JR (2013) Unsupervised adaptation for acceleration-based activity recognition: robustness to sensor displacement and rotation. Pers Ubiquitous Comput 17(3):479–490

Gjoreski H, Ciliberto M, Morales FJO, Roggen D, Mekki S, Valentin S (2017) A versatile annotated dataset for multimodal locomotion analytics with mobile devices. In: Proceedings of the 15th ACM conference on embedded network sensor systems, SenSys '17, New York, NY, USA, 2017. ACM, pp 61:1–61:2

Kunze K, Lukowicz P (2014) Sensor placement variations in wearable activity recognition. IEEE Pervasive Comput 13(4):32–41

Lara Oscar D, Labrador Miguel A (2013) A survey on human activity recognition using wearable sensors. IEEE Commun Surv Tutor 15(3):1192–1209

Lockhart JW, Weiss GM, Xue JC, Gallagher ST, Grosner AB, Pulickal TT (2011) Design considerations for the WISDM smart phone-based sensor mining architecture. In: Proceedings of the fifth international workshop on knowledge discovery from sensor data, SensorKDD '11, New York, NY, USA, 2011. ACM, pp 25–33

Micucci D, Mobilio M, Napoletano P (2017) UniMiB SHAR: a dataset for human activity recognition using acceleration data from smartphones. Appl Sci 7(10)

Ofli F et al (2013) Berkeley MHAD: a comprehensive multimodal human action database. In: 2013 IEEE WACV, pp 53–60

Roggen D, Calatroni A, Rossi M, Holleczek T, Förster K, Tröster G, Lukowicz P, Bannach D, Pirkl G, Ferscha A, Doppler J, Holzmann C, Kurz M, Holl G, Chavarriaga R, Sagha H, Bayati H, Creatura M, Millan JR (2010) Collecting complex activity datasets in highly rich networked sensor environments. In: 2010 seventh international conference on networked sensing systems (INSS), pp 233–240

Ruiz C, Pan S, Bannis A, Chen X, Joe-Wong C, Noh HY, Zhang P (2018) IDrone: robust drone identification through motion actuation feedback. Proc ACM Interact Mob Wearable Ubiquitous Technol 2(2):80:1–80:22

Stisen A, Blunck H, Bhattacharya S, Prentow TS, Kjærgaard MB, Dey A, Sonne T, Jensen MM (2015) Smart devices are different: assessing and mitigating mobile sensing heterogeneities for activity recognition. In: Proceedings of the 13th ACM conference on embedded networked sensor systems, SenSys '15, New York, NY, USA, 2015. ACM, pp 127–140

Takeda S, Lago P, Okita T, Inoue S (2018) A multi-sensor setting activity recognition simulation tool. In: Proceedings of the 2018 ACM international joint conference and 2018 international symposium on pervasive and ubiquitous computing and wearable computers, UbiComp '18, New York, NY, USA, 2018. ACM, pp 1444–1448

Ustev YE, Incel OD, Ersoy C (2013) User, device and orientation independent human activity recognition on mobile phones: challenges and a proposal. In: Proceedings of the 2013 ACM conference on pervasive and ubiquitous computing adjunct publication, UbiComp '13 Adjunct, New York, NY, USA, 2013. ACM, pp 1427–1436

Wüstenberg M, Lukowicz P, Kjaergaard MB, Blunck H, Grønbæk K, Franke T, Bouvin NO (2013) On heterogeneity in mobile sensing applications aiming at representative data collection. In: UbiComp '13 Adjunct proceedings of the 2013 ACM conference on pervasive and ubiquitous computing adjunct publication. ACM, pp 1087–1098

Yang J (2009) Toward physical activity diary: motion recognition using simple acceleration features with mobile phones. In: Proceedings of the 1st international workshop on interactive multimedia for consumer electronics, IMCE '09, New York, NY, USA. ACM, pp 1–10

Young AD, Ling MJ, Arvind DK (2011) IMUSim: a simulation environment for inertial sensing algorithm design and evaluation. In: Proceedings of the 10th ACM/IEEE IPSN, pp 199–210

Yurtman A, Barshan B, Fidan B (2018) Activity recognition invariant to wearable sensor unit orientation using differential rotational transformations represented by quaternions. Sensors 18(8)

Part III
SHL: An Activity Recognition Challenge

Many profound research questions can be formulated as a competitive event, a challenge, which then is posed year after year to the research community. This last part presents the latest challenge in activity recognition, the Sussex-Huawei Locomotion and transportation recognition challenge, as it was held in 2018. The five chapters in this part provide insights into the strategies that were followed and the open questions that have been revealed through this challenge.

In the first chapter of this book part, part of competition organizing team presents Chap. 11. The goal of the challenge is to recognize eight transportation activities (Still, Walk, Run, Bike, Bus, Car, Train, Subway) from the inertial and pressure sensor data of a smartphone. In this first chapter, a reference recognition performance obtained by applying various classical and deep-learning classifiers to the testing dataset, is presented. Results show that convolutional neural network operating on frequency-domain raw data achieves the best performance among all the classifiers.

The next Chap. 12, proposes a large-scale systematic experimental setup in order to design and evaluate neural architectures for activity recognition applications. The authors have trained and evaluated in this chapter more than 600 different architectures which are then analyzed to assess hyperparameters relevance, on the SHL challenge dataset.

In the 2018 Sussex-Huawei Locomotion-Transportation (SHL) recognition challenge, most submissions have strongly overestimated the performance of their algorithms in relation to their performance achieved on the challenge evaluation data. In the subsequent Chap. 13, an experiment demonstrates potential sources of upward scoring bias in the evaluation of travel activity classifiers. Three types of results are discussed, which can serve as guidelines to avoid the said overestimation of classifier performance.

Chapter 14 are discussed in the fourth chapter of Part III. In it, a data analysis pipeline that contains three stages, a preprocessing stage, a classification stage, and a time stabilization stage, is proposed. The first analysis shows promising results: Classification on a 10 hold-out sample of the training data on data features without feature extraction over extremely short time windows and then stabilizing the activity predictions over longer time windows results in a much higher accuracy. But in the second part of this chapter, this model is also found to not generalize well despite the use of a hold-out sample to prevent test set leakage.

Finally, the top winning submissions of the challenge is presented in the final Chap. 15. Both submissions start with data preprocessing, including a normalization of the phone orientation, to then handcrafted a set of domain features in both frequency and time domain. The second-place submission feeds the best features into an XGBoost machine-learning model with optimized hyperparameters, achieving the accuracy of 90.2

Chapter 11
Benchmark Performance for the Sussex-Huawei Locomotion and Transportation Recognition Challenge 2018

Lin Wang, Hristijan Gjoreski, Mathias Ciliberto, Sami Mekki, Stefan Valentin and Daniel Roggen

Abstract The Sussex-Huawei Transportation-Locomotion (SHL) Recognition Challenge 2018 aims to recognize eight transportation activities (Still, Walk, Run, Bike, Bus, Car, Train, Subway) from the inertial and pressure sensor data of a smartphone. In this chapter, we, as part of competition organizing team, present reference recognition performance obtained by applying various classical and deep-learning classifiers to the testing dataset. The classical classifiers include naive Bayes, decision tree, random forest, K-nearest neighbours and support vector machine, while the deep-

L. Wang (✉) · H. Gjoreski · M. Ciliberto · D. Roggen
Wearable Technologies Laboratory, Sensor Technology Research Centre, University of Sussex, Brighton, UK
e-mail: lin.wang@qmul.ac.uk

H. Gjoreski
e-mail: hristijang@feit.ukim.edu.mk

M. Ciliberto
e-mail: m.ciliberto@sussex.ac.uk

D. Roggen
e-mail: danie.roggen@icee.org

L. Wang
Centre for Intelligent Sensing, Queen Mary University of London, London, UK

H. Gjoreski
Faculty of Electrical Engineering and Information Technologies, Ss. Cyril and Methodius University in Skopje, Skopje, Macedonia

S. Mekki
Mathematical and Algorithmic Sciences Lab, PRC, Huawei Technologies France, Boulogne-Billancourt, France
e-mail: sami.mekki@huawei.com

S. Valentin
Department of Computer Science, Darmstadt University of Applied Sciences, Darmstadt, Germany
e-mail: stefan.valentin@h-da.de

N. Kawaguchi et al. (eds.), *Human Activity Sensing*,
Springer Series in Adaptive Environments,
https://doi.org/10.1007/978-3-030-13001-5_11

learning classifiers include fully-connected and convolutional deep neural networks. We feed different types of input to the classifier, including hand-crafted features, raw sensor data in the time domain, and in the frequency domain. We additionally employ a post-processing scheme, which smoothens the predictions in order and improves the recognition performance. Results show that convolutional neural network operating on frequency-domain raw data achieves the best performance among all the classifiers. Finally, we achieve a benchmark result with F1 score 92.9%, which is comparable to the best result from the team that won the competition (achieving F1 score 93.9%). The competition dataset and the benchmark implementation is made available online (http://www.shl-dataset.org/).

11.1 Introduction

Todays mobile phones come equipped with a rich set of sensors, including accelerometer, gyroscope, magnetometer, global positioning system (GPS), microphone, camera and others, which can be used to discover user activities and contex (Lane et al. 2010). Transportation mode is an important element of the users context that indicates how users move about, such as by walking, running, driving car or taking a bus (Engelbrecht et al. 2015). Transportation mode recognition is useful for a variety of applications, such as intelligent service adaptation, individual environmental impact monitoring, human-centered activity monitoring, and so on (Dobre and Xhafa 2014; Brazil and Caulfield 2013; Castignani et al. 2015).

There have been numerous studies showing how to utilize machine learning techniques to recognize transportation modes from multimodal smartphone sensors, such as inertial (accelerometer, gyroscope, and magnetometer) and GPS sensors (Hemminki et al. 2013; Xia et al. 2014; Yu et al. 2014; Fang et al. 2017; Feng and Timmermans 2013; Jahangiri and Rakha 2015; Su et al. 2016; Richoz et al. 2019). To date, most research groups assess the performance of their algorithms using their own collected data, which cover a different number of transportation activities and sensor modalities. Due to the complexity of the data collection procedure and the need to protect participant privacy, these ad-hoc datasets often have a short duration and remain private. This prevents the comparison of different approaches in a replicable and fair manner within and across research groups and impedes the progress in this research area (Wang et al. 2019).

To fill this gap, and to promote the advance of the research in the field, the Sussex-Huawei Locomotion-Transportation (SHL) recognition challenge was organized with a unified recognition task, dataset and sensor modalities (Gjoreski et al. 2018; Wang et al. 2018a, b). The objective is to recognize eight transportation activities (Still, Walk, Run, Bike, Bus, Car, Train, Subway) from the inertial and pressure sensor data of a smartphone. The challenge attracted 19 submissions internationally, and the competition outcome established a state-of-the-art performance in motion-based transportation recognition (with the highest F1 score 93.9%). In addition to the promising methodologies and results reported by the competition teams, the challenge also identified that motion sensors struggle distinguishing between transportation modes of similar classes: for example between train and subway (road transport) and between bus and car (road transport). For more details, please refer to the summary of the SHL challenge (Wang et al. 2018b).

In this chapter we, as part of the challenge organizing team, establish reference performance for the challenge by applying classical machine-learning and emerging deep-learning pipelines to the challenge dataset. Classical classifiers usually perform feature computation and classification independently. In a classical pipeline, hand-crafted features of the sensor data are first computed and their number is reduced through feature selection. This requires a deep understanding of the relationship between features and activities. Deep-learning pipelines instead learn the features and the classifier (deep neural network) simultaneously from the sensor data. It seamlessly integrates feature computation and classification and thus, theoretically, would not need additional interaction from researchers. Deep-learning pipelines have been employed in human activity recognition successfully (Ordonez and Roggen 2016), however their application in transportation mode recognition is still in a very early stage.

For classical pipelines we consider naive Bayes (NB), decision tree (DT), random forest (RF), k-nearest neighbours (KNN) and support vector machine (SVM), while for deep-learning pipelines we consider fully-connected deep neural network (FC) and convolutional neural network (CNN). We combine three inertial sensors (accelerometer, gyroscope and magnetometer), and feed various input to the the classifiers, including hand-crafted features, time-domain raw data and frequency-domain raw data. A post-processing scheme, which exploits the temporal correlation between neighbouring frames, is employed to further improve the recognition performance. We train the classifiers with the training dataset and report evaluation results on the testing dataset. Our benchmark result (F1 score 92.9%)is comparable with the best result achieved by the competing teams (F1 score 93.9%).

The chapter is originally based on the workshop paper (Wang et al. 2018a). In this extended version, we additionally include more technical details for implementing the recognition pipelines, in particular feature computation in the time and the frequency domain for classical pipeline. We make the competition dataset and the baseline code publicly available to encourage reproducible comparison.

The chapter is organized as below. Section 11.2 introduces the SHL dataset and the competition protocol. Section 11.3 describes the classical pipeline and deep-learning pipelines and Sect. 11.4 reports the benchmark results. Finally, Sect. 11.5 draws the conclusions.

11.2 Brief of the SHL Challenge

11.2.1 Dataset

The challenge uses a subset of the Sussex-Huawei Locomotion-Transportation (SHL) dataset (Gjoreski et al. 2018). The SHL dataset was recorded over a period of 7 months in 2017 by 3 participants engaging in 8 different modes of transportation in real-life setting in the United Kingdom, i.e. Still, Walk, Run, Bike, Car, Bus, Train, and Subway. Each participant carried four smartphones at four body positions simultaneously: in the hand, at the torso, in the hip pocket, in a backpack or handbag. The smartphone logged data from 16 sensor modalities (Ciliberto et al. 2017). The complete dataset contains up to 2812 h of labeled data, corresponding to 16,732 travel distance, and is considered as one of the biggest dataset in the research community.

The SHL recognition challenge uses the data recorded by the first participant with the phone at the hip pocket position, and includes 82 days of recording (5–8 h per day) during a 4-month period. The challenge uses 62 days as the training dataset (271 h) and 20 days for the testing dataset (95 h). Figure 11.1 depicts the duration of each transportation activity in the training and testing datasets. The dataset provides the raw data from 7 sensors, including accelerometer, gyroscope, magnetometer, linear acceleration, gravity, orientation, and ambient pressure. The sampling rate of all these sensors is 100 Hz.

The SHL recognition challenge used the data recorded by the first participant with the phone at the pocket position in 82 days (5–8 h per day). The challenge uses 62 days as the training dataset and 20 days for the testing dataset. The challenge dataset provides the raw data from accelerometer, gyroscope, magnetometer, linear accelerometer, gravity, orientation, and pressure. The sampling rate of all the sensors is $f_s = 100$ Hz.

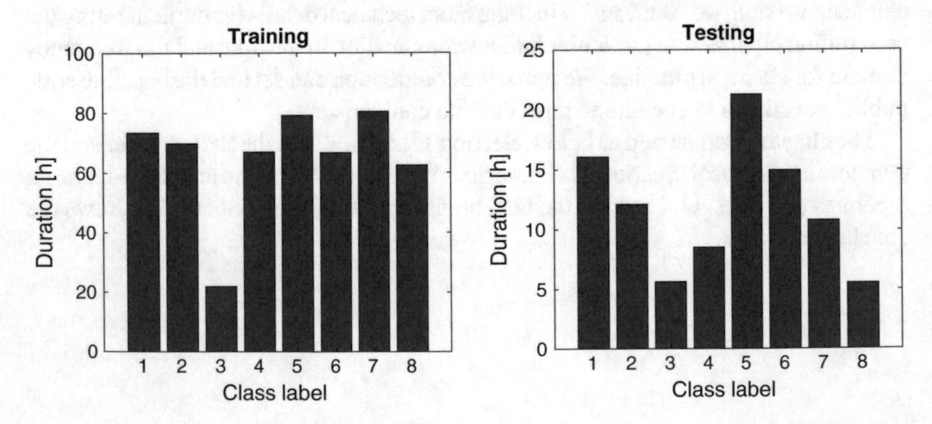

Fig. 11.1 The duration of each class activity in the training and the testing dataset. The 8 class activities are: 1—Still; 2—Walk; 3—Run; 4—Bike; 5—Car; 6—Bus; 7—Train; 8—Subway

11.2.2 Data Format

For both training and testing datasets, we chopped the data into segments with a 1-min non-overlap sliding window. The order of the segments was randomly permuted so that there was no temporal dependency among segments. This guarantees that the maximum frame size used by participants is one minute, and thus provides an upper bound on the latency of the recognition pipeline. For reference, the original order of segments in the training dataset is provided.

As shown in Table 11.1, the training set contains 21 plain text files (~5.5 GB) corresponding to various sensor channels, the label and the segment order. The testing set contains 19 plain text files (~1.9 GB), similar to the training dataset but without the label nor the segment order file.

Each sensor data file in the training set contains a matrix of size 16,310 lines × 6000 columns, corresponding to 16,310 segments each containing 6000 samples (1 min). The data in the label file is of the same size (16,310 × 6000), indicating sample-wise transportation activity. The 8 numbers in the label file indicate the 8 activities: 1—Still; 2—Walk; 3—Run; 4—Bike; 5—Car; 6—Bus; 7—Train; 8—Subway.

The testing set has the same structure as the training dataset, except that the data size is 5698 lines × 6000 columns, corresponding to 5698 segments each containing 6000 samples. The label file of the testing set will remain confidential until after the challenge. It is used for performance evaluation by the challenge organizer.

11.2.3 Task and Evaluation

The task is to train a recognition pipeline using the training dataset and then use this system to recognize the transportation mode from the sensor data in the testing set. The recognition performance is evaluated with the F1 score averaged over all the activities.

Let M_{ij} be the (i, j)-th element of the confusion matrix. It represents the number of samples originally belonging to class i which are recognized as class j. Let $C = 8$ be the number of classes. The F1 score is defined as below.

$$\text{recall}_i = \frac{M_{ii}}{\sum_{j=1}^{C} M_{ij}}, \quad \text{precision}_j = \frac{M_{jj}}{\sum_{i=1}^{C} M_{ij}}, \tag{11.1}$$

$$F1 = \frac{1}{C} \sum_{i=1}^{C} \frac{2 \cdot \text{recall}_i \cdot \text{precision}_i}{\text{recall}_i + \text{precision}_i}. \tag{11.2}$$

In this chapter, we additionally measure the global recognition accuracy over all the data samples.

Table 11.1 Data files provided by the SHL recognition challenge

Modality	File	Train	Test
Accelerometer	Acc_x.txt Acc_y.txt Acc_z.txt	✓	✓
Gyroscope	Gyr_x.txt Gyr_y.txt Gyr_z.txt	✓	✓
Magnetometer	Mag_x.txt Mag_y.txt Mag_z.txt	✓	✓
Linear accelerometer	LAcc_x.txt LAcc_y.txt LAcc_z.txt	✓	✓
Gravity	Gra_x.txt Gra_y.txt Gra_z.txt	✓	✓
Orientation	Ori_w.txt Ori_x.txt Ori_y.txt Ori_z.txt	✓	✓
Pressure	Pressure.txt	✓	✓
Label	Label.txt	✓	×
Order	train_order.txt	✓	×

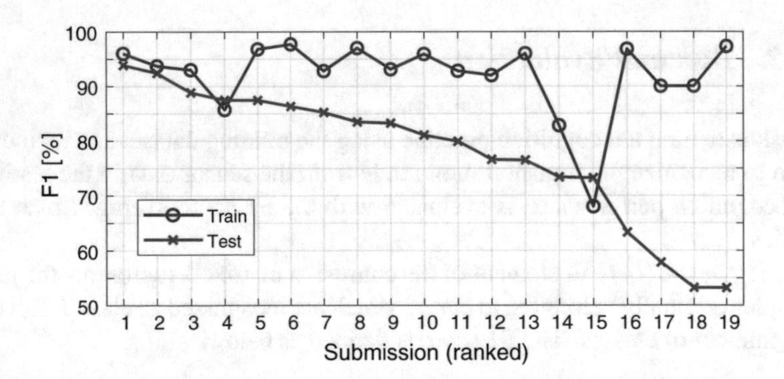

Fig. 11.2 F1 scores obtained by the SHL competing teams for the training and testing datasets

The challenge has attracted 19 submissions with the results shown in Fig. 11.2. Note that the benchmark result is not included in the official competition and thus is not shown in this figure.

11.3 Method

11.3.1 (a) Processing Pipeline

Figure 11.3 depicts the processing pipeline for predicting the transportation mode from the multimodal sensor data. For the training dataset, the sensor data from M modalities are segmented into L_t short frames: $\{\mathbf{s}_t(1), \ldots, \mathbf{s}_t(l), \ldots, \mathbf{s}_t(L_t)\}$, where $\mathbf{s}_t(l)$ represents the sensor data in the l-th frame. Given the labels in these frames, $\{c_t(1), \ldots, c_t(l), \ldots, c_t(L_t)\}$, are known, the training data is used to train a classifier model. Subsequently, the sensor data from testing dataset is segmented into L frames: $\{\mathbf{s}(1), \ldots, \mathbf{s}(L)\}$, which are mapped into one of the transportation classes with the trained classifier mode. A post-processor follows to improve the recognition results at individual frames.

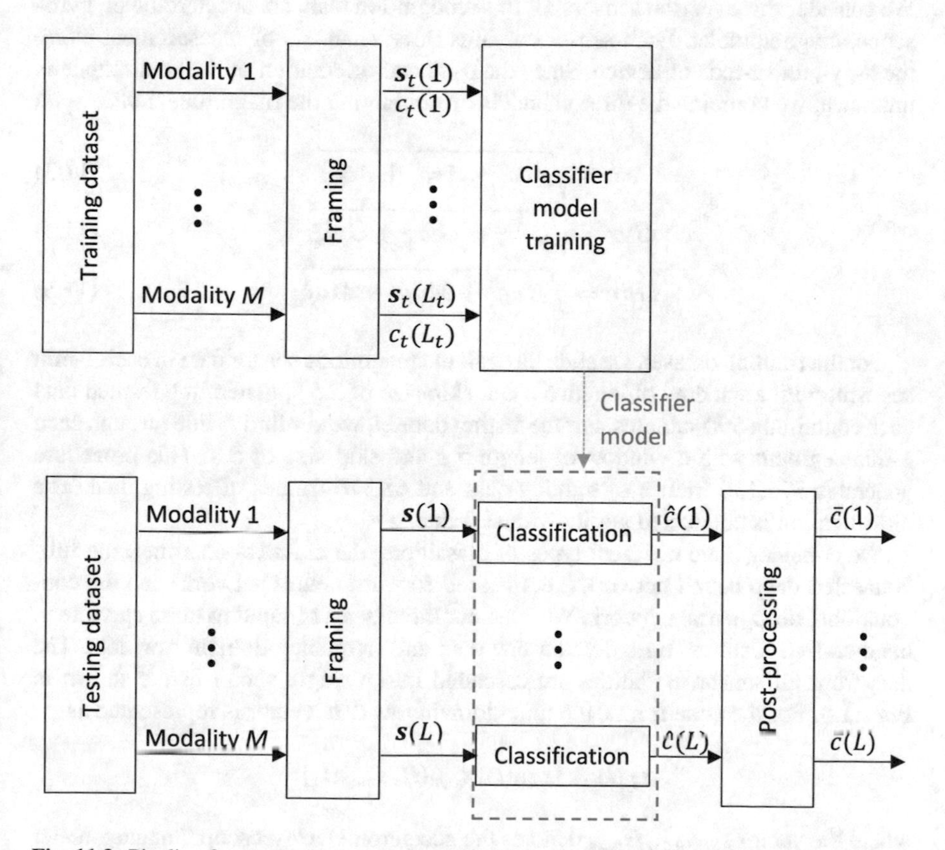

Fig. 11.3 Pipeline for transportation mode recognition using the training and the testing datasets of the SHL recognition challenge

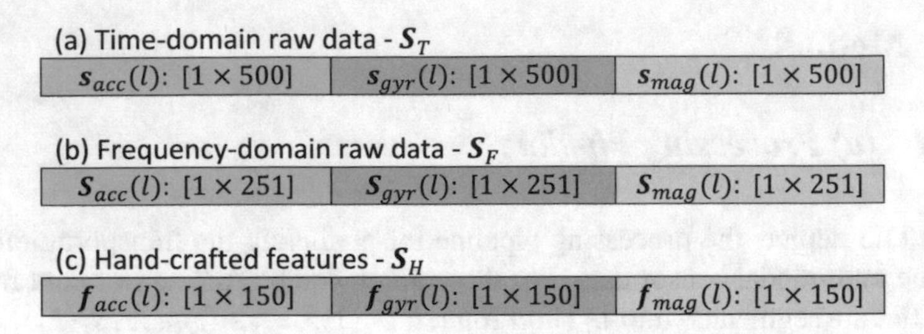

Fig. 11.4 Three types of input data to the classifier cascading three sensor modalities. **a** Time-domain raw data. **b** Frequency-domain raw data. **c** Hand-crafted features

11.3.2 (b) Sensor Input

We consider three inertial sensors for this recognition task, i.e. accelerometer, gyroscope, magnetometer. Each sensor contains three channels of measurement along the x-, y-, and z-axis of device. Since the pose and orientation of the smartphone is unknown, we combine the three channels by computing the magnitude, i.e.

$$Acc = \sqrt{Acc_x^2 + Acc_y^2 + Acc_z^2} \tag{11.3}$$

$$Gyr = \sqrt{Gyr_x^2 + Gyr_y^2 + Gyr_z^2} \tag{11.4}$$

$$Mag = \sqrt{Mag_x^2 + Mag_y^2 + Mag_z^2} \tag{11.5}$$

For the training dataset, we slide through the magnitude sensor data in each 1-min segment with a window of length 5 s and skip size of 2.5 s, generating framed data each containing 500 samples. For the testing dataset, we similarly slide through each 1-min segment with a window of length 5 s and skip size of 5 s. This procedure generates 375,130 frames of training data and 68,376 frames of testing data. The classification is conducted per individual frame.

We consider three different types of classifiers, the classical classifier, the full-connected deep neural network (i.e. the feed-forward neural network) and the convolutional deep neural network. We consider three types of input to these classifiers: hand-crafted features, time-domain raw data and frequency-domain raw data. The data from all sensor modalities are cascaded into a single vector as the shown in Fig. 11.4. For the l-the frame, the time-domain raw data vector is represented as

$$\boldsymbol{s}_T(l) = [\boldsymbol{s}_{acc}(l), \boldsymbol{s}_{gyr}(l), \boldsymbol{s}_{mag}(l)],$$

where the vector $\boldsymbol{s}_{acc}/\boldsymbol{s}_{gyr}/\boldsymbol{s}_{mag}$ denotes the accelerometer/gyroscope/magnetometer magnitude samples in the l frame.

The frequency-domain raw data vector is represented as

$$\boldsymbol{s}_F(l) = [\boldsymbol{S}_{acc}(l), \boldsymbol{S}_{gyr}(l), \boldsymbol{S}_{mag}(l)],$$

where $\boldsymbol{S}_{acc}/\boldsymbol{S}_{gyr}/\boldsymbol{S}_{mag}$ denotes the magnitude of FFT version of $\boldsymbol{s}_{acc}/\boldsymbol{s}_{gyr}/\boldsymbol{s}_{mag}$ (retaining frequencies $[0, fs/2]$).

The feature vector is represented as

$$\boldsymbol{s}_H(l) = [\boldsymbol{f}_{acc}(l), \boldsymbol{f}_{gyr}(l), \boldsymbol{f}_{mag}(l)],$$

where $\boldsymbol{f}_{acc}/\boldsymbol{f}_{gyr}/\boldsymbol{f}_{mag}$ denotes the hand-crafted features computed on the accelerometer/gyroscope/magnetometer sequence in the l-th frame. For each sensor modality, we compute 150 features per frame, as suggested in (Wang et al. 2019) (see Appendix for the definition of the features). This feature set is a hybrid combination of statistical features from the time and the frequency domain, the quantile range of the data value, and the subband energy, etc.

The time-domain raw data and the frequency-domain raw are mainly used as input to the deep-learning pipeline while the hand-crafted features are used for both classical and deep-learning pipelines. In the following, we discuss the details of the three types of classifiers.

11.3.3 (c) Classical Pipeline

Figure 11.5 illustrates the pipeline for predicting the transportation mode with a classical classifier, which uses as input the hand-crafted feature vector $\boldsymbol{s}_H$. A normalization procedure, which maps each feature into the range [0, 1], is applied before feeding the features to the classifier. This step aims to increase the robustness of the classifier.

Let's use the i-th feature f_i as an example. Suppose Q_i^{95} and Q_i^5 are the quantile 95 and quantile 5 of f_i across all the frames in the training dataset. The normalization of each frame is performed as

$$\bar{f}_i(l) \leftarrow \min\left(\max\left(\frac{f_i(l) - Q_i^5}{Q_i^{95} - Q_i^5}, 0\right), 1\right). \tag{11.6}$$

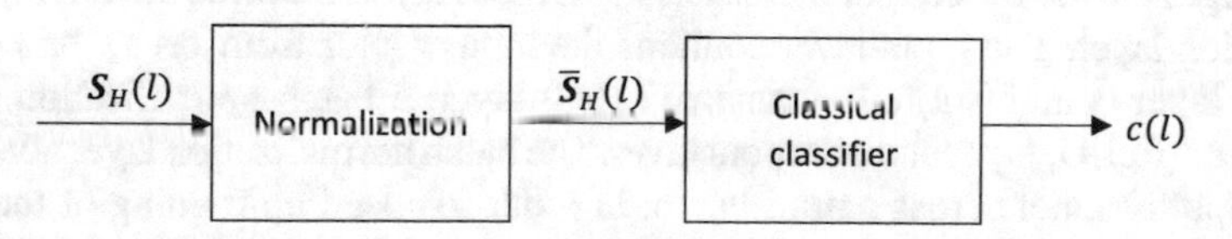

Fig. 11.5 The processing pipeline using a classical classifier

Table 11.2 Configuration of the classical classifiers. We use default parameters of the Matlab function unless explicitly mentioned

Classifier	Matlab function	Parameter
NB	fitcnb	/
DT	fitcdt	minleafsize = 1000
RF	TreeBagger	NumTrees = 20 minleafsize = 1000
KNN	fitcknn	/
SVM	svmtrain svmpredict	/

The features in the testing dataset are normalized in the same way using Q_i^{95} and Q_i^5, which are computed in advance from the training dataset. After normalization, the new feature vector in the l-th frame is denoted as $\bar{s}_H(l)$.

Table 11.2 lists the five classical classifiers that are considered for the recognition task: NB, DT, RF, KNN and SVM. The first four classifiers are implemented with Matlab Machine Learning Toolbox, while SVM is implemented with LIBSVM (Chang and Lin 2011). All the classifiers use default parameters set in the library functions unless explicitly mentioned. For instance, the KNN Matlab function uses a default value of $K = 1$. We can also set the parameters manually. For instance, in DT we set the parameter 'minleafsize = 1000'; in RF we set the number of trees as 20 and in each tree the parameter 'minleafsize = 1000'.

11.3.4 (d) Fully-Connected Deep Neural Network

Here a fully-connected deep neural network (FC) refers to a feed-forward neural network (we use this name throughout the chapter for consistence with Matlab Deep Learning library functions).

Figure 11.6 illustrates the architecture of the FC, which predicts the transportation mode using one of the three types of input: time-domain raw data ($\boldsymbol{s}_T$), frequency-domain raw data ($\boldsymbol{s}_F$) or hand-crafted features ($\boldsymbol{s}_H$). Similar to Eq. (11.6), all the inputs are normalized into the range [0, 1] before going to the classifier.

In Fig. 11.6 the FC classifier consists of an input layer, multiple hidden layers and a decision layer. The input layer contains the input vector from $\boldsymbol{s}_T$, $\boldsymbol{s}_F$ or $\boldsymbol{s}_H$. Each hidden layer contains a fully-connected (FC) layer, a batch normalization layer, a nonlinear (ReLU) layer and a dropout layer. The batch normalization layer normalizes each input channel across a mini-batch, in order to speed up training of the neural network and to reduce the sensitivity to network initialization. The dropout layer randomly sets input elements to zero with a given probability in order to prevent overfitting (Srivastava et al. 2014). The decision layer contains a full-connected layer, a nonlinear (SoftMax) layer and a classification layer, which finally infers the

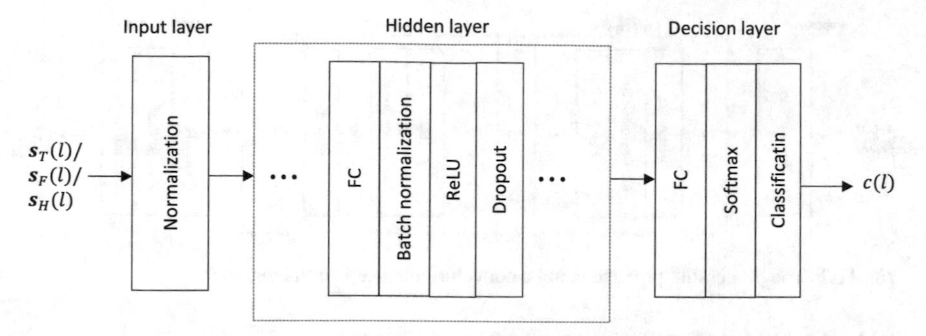

Fig. 11.6 The processing pipeline using a fully-connected deep neural network

Table 11.3 Configuration of the fully-connected deep neural network

Input layer	$s_T/s_F/s_H$
Hidden layer	Number of layers = 3
	Number of nodes per layer = 500
	Dropout ratio = 25%
Mini-batch size	500

transportation mode of the current frame. Due to the large amount of data, we employ a mini-batch processing scheme which updates the weights of the neural network per subset of training samples.

We use the Matlab Deep Learning Toolbox to implement the FC classifier. Table 11.3 shows the parametric configuration of the FC classifier.

11.3.5 (e) Convolutional Deep Neural Network

Figure 11.7 illustrates the architecture of a convolutional deep neural network (CNN), which predicts the transportation mode using one of the two types of input: time-domain raw data (s_T), frequency-domain raw data (s_F). Similar to Eq. (11.6), all the inputs are normalized into the range [0, 1] before going to the classifier.

In Fig. 11.7 the CNN classifier consists of an input layer, multiple CNN layers, multiple FC layers and a decision layer. The CNN layer contains multiple hidden layers, where each layer consists of a convlutional layer, a batch normalization layer and a nonlinear (ReLU) layer. The FC layer contains multiple hidden layers, where each layer consists of a fully-connected layer, a batch normalization layer, a nonlinear (ReLU) layer and a dropout layer. The decision layer consists of a fully-connected layer, a nonlinear (Softmax) layer and a classification layer which infers the transportation mode of the current frame. A mini-batch processing scheme is employed to deal with the large amount of data.

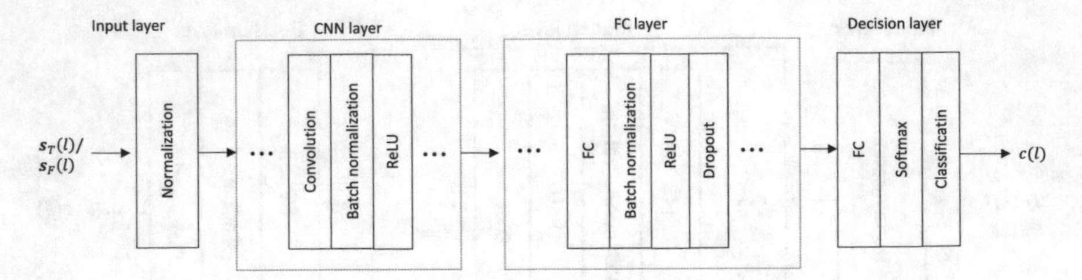

Fig. 11.7 The processing pipeline using a convolutional deep neural network

Table 11.4 Configuration of the convolutional neural network

Input layer	s_T/s_F
Convolution layer	Number of layers = 3
	Stride/kernal size per layer = 15
	Padding size per layer = 0
	Number of channels per layer = 100
FC hidden layer	Number of layers = 3
	Number of nodes per layer = 300
	Dropout ratio = 25%
Mini-batch size	500

We use the Matlab Deep Learning Toolbox to implement the CNN classifier. Table 11.4 shows the parametric configuration of CNN classifier.

11.3.6 *(f) Post-processing*

The classical and deep-learning systems usually make a decision per frame (5 s). In the SHL recognition challenge, the data was chopped into segments of 1 min long. Since the transportation mode of a user typically continues for a certain period and there is a strong correlation between neighbouring frames (Zhang and Poslad 2013; Yu et al. 2014), we reasonably assume that the transportation mode remains the same in the 1-min segment. Based on this assumption, we propose a majority voting scheme to further improve the recognition performance at individual frames.

Suppose the prediction results for the F frames in the s-th segment are $c_s(1), \ldots, c_s(F)$, and the occurrence of each activity is denoted as $N_s(1), \ldots, N_s(8)$. The transportation mode in this segment is unified as

$$\bar{c}_s = \max_{i \in [1,8]} N_s(i), \tag{11.7}$$

and the prediction of all frames is updated as

$$\hat{c}_s(f) = \bar{c}_s, \quad f = 1, \ldots, F \tag{11.8}$$

11.4 Results

For data processing we use a computer equipped with an Intel i7-4770 4-core CPU @ 3.40 GHz with 32 GB memory, and a GeForce GTX 1080 Ti GPU with 3584 CUDA cores @ 1.58 GHz and 11 GB memory. The code is written with Matlab 2018a, calling functions from the Machine Learning Toolbox and the Deep Learning Toolbox.

As indicated in Fig. 11.3, we use the training dataset (375,130 frames) to train the classifier, and then use the testing dataset (68,376 frames) to evaluate the performance. Table 11.5 summarizes the global accuracy and the F1 score before and after post-processing for each classifier, and the processing time for training and testing. The two measures, global accuracy and F1 score, do not show big difference for most classifiers.

For each classifier, post-processing can improve the recognition performance effectively (F1 score by above 10%). However, for some classifiers with low recognition accuracy before post-processing, the improvement is not so evident, e.g. NB and FC-time.

CNN-frequency performs the best among all the candidates with the second-highest F1 score before post-processing and the highest F1 score (**92.9%**) after post-processing. FC-frequency achieves the third highest F1 score before post-processing and the second highest F1 score after post-processing. FC-feature achieves the highest F1 score (**82.7%**) before post-processing and the third highest F1 score after post-processing.

For the same input (e.g. time-domain or frequency-domain raw data), CNN-based classifiers usually outperform FC-based classifiers. In particular, for the two classifiers which both operate on the time-domain raw data, CNN-time significantly outperforms FC-time. Frequency-domain raw data tends to provide more insightful information than time-domain raw data, for both FC and CNN. Deep-learning classifiers also work well with hand-crafted features, with FC-feature outperforming all the five classical classifiers. Among the five classical classifiers, SVM achieves the highest F1 score (before and after post-processing) while KNN performs the second best. RF performs slightly better than DT, while NB performs the worst.

SVM is the most time-consuming algorithm for both training (32,418 s) and testing (7777 s) among the ten classifiers. It takes even longer training time than the deep-learning classifiers. KNN takes the least time for training (1.3 s) among all the ten classifiers but takes the second longest time for testing (2567 s). KNN and SVM

Table 11.5 Evaluation results on the testing dataset of the SHL recognition challenge. KEY: A—accuracy; F1—F1 score; PP—post-processing; PT—processing time

	Before PP		After PP		PT (s)	
Classifier	A (%)	F1 (%)	A (%)	F1 (%)	Train	Test
NB	63.7	59.1	69.8	62.0	13	1.1
DT	71.2	72.6	80.7	82.0	91	0.1
RF	76.5	76.6	84.5	84.5	91	2.5
KNN	70.4	71.3	85.8	85.3	1.3	2567
SVM	78.7	79.2	87.1	87.0	324,18	7777
FC-time	68.1	68.7	71.4	71.1	427	2.3
FC-frequency	82.5	81.7	91.5	90.4	362	2.0
FC-feature	82.4	**82.7**	89.9	90.2	502	3.1
CNN-time	80.3	80.8	85.9	86.6	8122	14.9
CNN-frequency	**83.0**	82.5	**93.3**	**92.9**	4604	7.5

are the two classifiers that take over 2000 s for testing. DT and RF both take about 90 s for training and less than 3 s for testing, while DT takes the least testing time (0.1 s) among all the ten classifiers. The long testing time by SVM is quite surprising because the classification procedure of SVM is a linear operation, which is very fast.

Deep-learning classifiers usually take hundreds to thousands of seconds for training. CNN-based classifiers take 10 times longer than FC-based classifiers for training. Deep-learning classifiers take several seconds for testing, which is comparable to classical classifiers. CNN-time takes the longest time for testing (14.9 s) among all the five deep-learning classifiers.

Table 11.6 presents the confusion matrices obtained by the 10 classifiers. The confusion matrices show that the first four activities (Still, Walk, Run, and Bike) are better recognized compared to the last four (Car, Bus, Train, and Subway). The motion of the smartphones during walk, run and bike is significantly higher than when the person is sitting or standing in the car, bus, train or subway, thus making the former four more distinctive than the latter four. There is mutual confusion between the motor vehicles (Car vs. Bus), and between the rail vehicles (Train vs. Subway). The reason for this is the similar motion patterns during these activities. Some confusion between Still and the four vehicle activities (Car, Bus, Train and Subway) is also observed. Typically, some vehicle classes are recognized as Still. This is possibly because the smartphones tend to be motionless when vehicle stops. Post-processing can improve the recognition accuracy for every class activity.

Table 11.6 Confusion matrices obtained by the considered classifiers. The F1 scores before(after) post-processing are also given. The 8 class activities are: 1—Still; 2—Walk; 3—Run; 4—Bike; 5—Car; 6—Bus; 7—Train; 8—Subway. Key: PP—post-processing

NB F1=59.1%(62.0%)

PP	Groundtruth class	1	2	3	4	5	6	7	8
Before PP	1	84	0	0	1	8	1	3	2
	2	2	79	8	8	0	0	0	2
	3	0	4	96	0	0	0	0	0
	4	2	30	7	58	1	2	0	1
	5	3	0	0	0	76	9	7	4
	6	9	0	0	2	57	16	8	8
	7	6	0	0	1	11	1	72	9
	8	4	0	0	0	5	0	80	10
After PP	1	95	0	0	0	3	0	1	0
	2	3	86	5	5	0	0	0	0
	3	0	3	96	0	0	0	0	0
	4	1	33	7	58	0	0	0	0
	5	2	0	0	0	90	4	3	1
	6	6	1	0	0	75	11	3	5
	7	6	1	0	0	4	0	85	3
	8	2	0	0	0	1	0	93	4

DT F1=72.7%(82.0%)

PP	Groundtruth class	1	2	3	4	5	6	7	8
Before PP	1	90	1	0	0	1	4	2	2
	2	4	89	0	4	0	1	0	2
	3	0	2	97	1	0	0	0	0
	4	2	8	1	86	1	3	0	0
	5	3	0	0	1	53	37	4	2
	6	7	1	0	2	18	62	5	5
	7	7	1	0	0	6	8	53	24
	8	7	1	0	0	2	1	32	57
After PP	1	97	0	0	0	0	1	1	0
	2	4	95	0	0	0	0	0	1
	3	0	1	98	0	0	0	0	0
	4	2	2	0	96	0	0	0	0
	5	2	0	0	0	63	33	1	0
	6	5	1	0	1	13	77	1	2
	7	7	1	0	0	3	7	67	15
	8	5	0	0	0	1	0	28	66

RF F1=76.6%(84.5%)

PP	Groundtruth class	1	2	3	4	5	6	7	8
Before PP	1	92	1	0	1	2	2	2	2
	2	3	92	0	2	0	0	0	1
	3	0	3	96	0	0	0	0	0
	4	2	7	2	87	1	1	0	0
	5	3	0	0	0	69	25	2	1
	6	5	1	0	1	20	65	4	4
	7	7	1	0	1	5	7	57	23
	8	7	0	0	0	1	1	32	59
After PP	1	98	0	0	0	1	0	.0	0
	2	3	95	0	1	0	0	0	0
	3	0	3	97	0	0	0	0	0
	4	1	4	1	94	0	0	0	0
	5	2	0	0	0	79	18	0	0
	6	4	1	0	0	16	77	1	1
	7	8	1	0	0	3	4	69	16
	8	5	0	0	0	0	0	27	68

KNN F1=71.3%(85.3%)

PP	Groundtruth class	1	2	3	4	5	6	7	8
Before PP	1	79	3	0	3	6	2	5	3
	2	3	92	1	1	0	1	1	1
	3	0	1	98	0	0	0	0	0
	4	2	5	1	90	1	1	0	0
	5	5	0	0	1	61	24	5	3
	6	7	1	0	2	19	60	7	4
	7	6	1	0	1	9	9	47	27
	8	6	1	0	0	7	3	33	50
After PP	1	97	1	0	0	1	1	0	0
	2	2	98	0	0	0	0	0	0
	3	0	1	99	0	0	0	0	0
	4	1	3	0	96	0	0	0	0
	5	3	0	0	0	82	13	1	0
	6	4	1	0	0	12	80	2	1
	7	7	1	0	0	4	5	68	16
	8	4	1	0	0	3	0	30	62

SVM F1=79.2% (87.0%)

PP	Groundtruth class	1	2	3	4	5	6	7	8
Before PP	1	92	1	0	0	1	2	2	1
	2	3	95	0	0	0	0	0	1
	3	0	2	98	0	0	0	0	0
	4	2	3	0	93	0	1	0	0
	5	4	0	0	0	69	23	2	2
	6	5	1	0	1	17	70	3	3
	7	7	1	0	0	4	7	59	22
	8	6	1	0	0	2	1	29	62
After PP	1	98	0	0	0	0	0	0	0
	2	3	97	0	0	0	0	0	0
	3	0	1	99	0	0	0	0	0
	4	1	1	0	97	0	0	0	0
	5	2	0	0	0	81	15	0	1
	6	3	1	0	0	12	82	0	1
	7	7	1	0	0	1	4	73	14
	8	4	1	0	0	0	0	26	70

FC-time F1=68.7%(71.6%)

PP	Groundtruth class	1	2	3	4	5	6	7	8
Before PP	1	88	1	0	1	2	5	2	1
	2	3	94	0	1	0	0	0	1
	3	0	1	98	1	0	0	0	0
	4	2	2	1	94	0	1	0	0
	5	5	0	0	0	62	27	5	1
	6	23	1	0	1	32	36	3	3
	7	9	1	0	1	10	7	48	24
	8	5	1	0	0	2	5	55	32
After PP	1	93	0	0	0	0	4	1	0
	2	2	97	0	0	0	0	0	0
	3	0	1	99	0	0	0	0	0
	4	1	1	0	98	0	0	0	0
	5	4	0	0	0	69	24	2	0
	6	26	1	0	0	36	34	1	2
	7	9	1	0	0	7	5	53	25
	8	2	0	0	0	2	3	62	31

FC-frequency F1=81.7%(90.4%)

PP	Groundtruth class	1	2	3	4	5	6	7	8
Before PP	1	91	2	0	1	2	1	1	2
	2	3	94	0	1	0	0	0	2
	3	0	2	97	1	0	0	0	0
	4	2	6	1	90	0	1	0	0
	5	4	0	0	0	81	13	2	1
	6	4	1	0	1	10	78	3	3
	7	9	1	0	1	3	3	61	23
	8	7	1	0	0	2	1	24	66
After PP	1	97	1	0	1	1	0	0	1
	2	2	97	0	0	0	0	0	0
	3	0	1	98	0	0	0	0	0
	4	1	3	0	96	0	0	0	0
	5	2	0	0	0	94	4	0	0
	6	3	1	0	0	4	90	0	1
	7	8	1	0	0	1	1	74	14
	8	4	1	0	0	1	0	17	77

FC-feature F1=82.7%(90.2%)

PP	Groundtruth class	1	2	3	4	5	6	7	8
Before PP	1	92	2	0	1	1	1	2	1
	2	3	95	0	0	0	0	0	1
	3	0	2	98	0	0	0	0	0
	4	2	1	0	96	0	0	0	0
	5	1	0	0	0	72	24	2	1
	6	2	1	0	1	12	81	2	3
	7	6	1	0	0	3	4	64	21
	8	6	1	0	0	1	1	23	69
After PP	1	98	1	0	0	0	0	1	0
	2	3	97	0	0	0	0	0	0
	3	0	1	98	0	0	0	0	0
	4	2	1	0	97	0	0	0	0
	5	0	0	0	0	82	17	0	0
	6	1	1	0	0	6	90	0	1
	7	6	1	0	0	1	3	78	12
	8	3	1	0	0	0	0	12	84

CNN-time F1=80.8%(86.6%)

PP	Groundtruth class	1	2	3	4	5	6	7	8
Before PP	1	92	2	0	1	1	1	2	1
	2	3	95	0	0	0	0	1	1
	3	0	1	99	0	0	0	0	0
	4	2	1	0	96	0	0	0	0
	5	1	0	0	0	64	33	1	1
	6	2	0	0	1	8	85	2	2
	7	7	1	0	0	4	9	59	20
	8	7	1	0	0	0	1	31	60
After PP	1	97	1	0	0	0	1	1	0
	2	2	97	0	0	0	0	0	0
	3	0	1	99	0	0	0	0	0
	4	1	0	0	98	0	0	0	0
	5	1	0	0	0	70	29	0	0
	6	1	1	0	0	4	93	0	1
	7	7	1	0	0	3	8	72	10
	8	5	0	0	0	0	0	25	69

CNN-frequency F1=82.5%(92.9%)

PP	Groundtruth class	1	2	3	4	5	6	7	8
Before PP	1	89	3	0	1	1	1	4	2
	2	3	94	0	1	0	0	0	1
	3	0	1	98	1	0	0	0	0
	4	2	4	1	91	0	1	0	0
	5	2	0	0	0	75	19	2	1
	6	2	1	0	1	6	85	2	2
	7	6	1	0	1	2	5	67	18
	8	5	1	0	0	1	1	27	64
After PP	1	97	1	0	0	0	0	1	0
	2	1	98	0	0	0	0	0	0
	3	0	1	99	0	0	0	0	0
	4	1	2	0	97	0	0	0	0
	5	0	0	0	0	91	8	1	0
	6	1	1	0	0	2	95	0	0
	7	6	1	0	0	1	3	84	6
	8	2	1	0	0	0	0	17	81

Predicted class

11.5 Conclusion

We compared the transportation mode recognition performance obtained by classical and deep-learning pipelines, with input from hand-crafted features, time-domain raw data and frequency-domain raw data of the SHL challenge dataset. Unsurprisingly, random forest is highly recommend among the five classical classifiers, trading off the classification performance and the computational time. Deep-learning classifiers achieve better performance than classical ones at the cost of longer training time. CNN-based classifier operating on frequency-domain raw data achieves the best performance among all the classifiers. The post-processing scheme can improve the recognition performance remarkably.

The chapter mainly aims to provide a reference performance for comparison with the results from challenge participants (which will be presented in Wang et al. 2018b). We used off-the-shelf software with default setting to implement the recognition pipeline. There should be a lot of space to optimize the performance and the processing speed. For instance, the parameters of the classifier (in particular the classical ones) can be tuned trading off under-fitting and over-fitting. The slow classification time of the SVM library seems to be very odd. A thorough theoretical analysis to explain the better performance of the deep-learning classifiers would be our future work. The competition dataset and the benchmark implementation is made available online (http://www.shl-dataset.org/).

Appendix—Hand-Crafted Features

In the appendix we give the definition of hand-crafted features as proposed in (Wang et al. 2019). We compute the feature for the three sensor modalities: accelerometer, gyroscope and magnetometer. For each modality we use the magnitude of the data channel for feature computation. The magnitude has been widely used in the literature and is robust to the variation of device orientation. We compute the features within a short-time window of 5 s length and 2.5 s overlap. For each modality, we compute the same set of features as shown in Table 11.7. The features to be computed can be categorized into three families: subband energy ($\mathcal{E}$), time-domain quantile ($\mathcal{Q}$), and the remaining time-domain and frequency-domain ($\mathcal{T}+\mathcal{F}$) features.

We compute 17 time-frequency features containing 8 elements in the time domain and 9 in the frequency domain. These time-domain ($\mathcal{T}$) and frequency-domain ($\mathcal{F}$) features are the ones that are most popularly used in the literature.

Table 11.7 Hand-crafted features: subband ($\mathcal{E}$) and quantile ($\mathcal{Q}$) features, and the remaining time-frequency domain ($\mathcal{T}+\mathcal{F}$) features

Type	Features	Dimension
$\mathcal{E}$	Energy and energy ratio with scan width 1 Hz and skip 0.5 Hz	198
	Energy and energy ratio with scan width 2 Hz and skip 1 Hz	98
	Energy and energy ratio with scan width 3 Hz and skip 1 Hz	96
	Energy and energy ratio with scan width 4 Hz and skip 1 Hz	94
	Energy and energy ratio with scan width 5 Hz and skip 1 Hz	92
	Energy and energy ratio with scan width 10 Hz and skip 1 Hz	82
	Energy and energy ratio with scan width 15 Hz and skip 1 Hz	72
	Energy and energy ratio with scan width 20 Hz and skip 1 Hz	62
	Energy and energy ratio with scan width 25 Hz and skip 1 Hz	52
Total		**846**
$\mathcal{Q}$	Quartiles: [0, 5, 10, 25, 50, 75, 90, 95, 100]	9
	Pairwise quartile range for the 9 quartiles	36
Total		**45**
$\mathcal{T}$	Mean, standard deviation, energy	3
	Mean crossing rate	1
	Kurtosis and Skewness	2
	Highest auto correlation value and offset	2
$\mathcal{F}$	DC component of FFT	1
	Highest FFT value and frequency	2
	Ratio between the highest and the second FFT peaks	1
	Mean, standard deviation	2
	Kurtosis and skewness	2
	Energy	1
Total		**17**

The subband is defined with two parameters: centre frequency ω_c and bandwidth ω_b. The frequencies in a subband is thus given by $\omega \in [\omega_c - \frac{\omega_b}{2}, \omega_c + \frac{\omega_b}{2}]$. We compute a set of subband features with all possible parameters of ω_c and ω_b. The highest frequency of the data is 50 Hz as the sampling rate is 100 H. We consider the following bandwidth: $\omega_b \in \{1, 2, 3, 4, 5, 10, 15, 20, 25\}$ Hz. For each bandwidth ω_b, we vary the centre frequency from $\frac{\omega_b}{2}$ to $50 - \frac{\omega_b}{2}$ with a step of 1 Hz. For the bandwidth $\omega_b = 1$ Hz the center frequency is increased with a step of 0.5 Hz. For each subband, we consider two types of features: the absolute energy and the energy ratio. Let $\{S_1, \ldots, S_K\}$ represent the $K = 257$ FFT coefficients of a frame of data and let k_L and k_H denotes the indices of the lower and upper frequencies of a subband $[\omega_c - \frac{\omega_b}{2}, \omega_c + \frac{\omega_b}{2}]$, the two features are defined as

$$f_{\text{subegr}} = \sum_{k=k_L}^{k_H} |S_k|^2, \tag{11.9}$$

$$f_{\text{subratio}} = \frac{\sum_{k=k_L}^{k_H} |S_k|^2}{\sum_{k=1}^{K} |S_k|^2}. \tag{11.10}$$

Finally we obtain 846 features in the set $\mathcal{E}$.

The quantile range $[q_L, q_H]$ is defined as $s(q_H) - s(q_L)$, the difference between two percentile values $s(q_L)$ and $s(q_H)$ of a frame of samples s. We compute a set of quantile features with a list of possible parameters of q_L and q_H. We consider the following 9 quantile values $q_L, q_H \in \{0, 5, 10, 25, 50, 75, 90, 95, 100\}$ with $q_L \leq q_H$. This results in 9 quantiles with $q_L = q_H$ and 36 quantile ranges with $q_L < q_H$. Finally we obtain 45 features in the set $\mathcal{Q}$.

As shown in Table 11.7, we in total compute $17 + 45 + 855 = 908$ features for each modality and thus $3 \times 908 = 2724$ features per frame of inertial sensor data in total. We further apply a filter-based feature selection algorithm employing a maximum-relevance-minimum-redundancy (MRMR) criteria (Peng et al. 2005) to select a subset of features. For each modality, we select about 150 featuers. While the featuers are selected using the full SHL dataset (Wang et al. 2019), we still use it for the challenge data. The computation of the features and the index of the selected ones will be given together with the code.

Acknowledgements This work was supported by the HUAWEI Technologies within the project "Activity Sensing Technologies for Mobile Users".

References

Brazil W, Caulfield B (2013) Does green make a difference: the potential role of smartphone technology in transport behaviour. Transp Res Part C Emerg Technol 37:93–101

Castignani G, Derrmann T, Frank R, Engel T (2015) Driver behavior profiling using smartphones: a low-cost platform for driver monitoring. IEEE Intell Transp Syst Mag 7(1):91–102

Chang CC, Lin CJ (2011) LIBSVM: a library for support vector machines. ACM Trans Intell Syst Technol 2(3):1–27

Ciliberto M, Ordonez FJ, Gjoreski H, Mekki S, Valentin S, Roggen D (2017) High reliability Android application for multidevice multimodal mobile data acquisition and annotation. In: Proceedings of ACM conference on embedded networked sensor systems, pp 1–2

Dobre C, Xhafa F (2014) Intelligent services for Big Data science. Futur Gener Comput Syst 37:267–281

Engelbrecht J, Booysen MJ, van Rooyen GJ, Bruwer FJ (2015) Survey of smartphone-based sensing in vehicles for intelligent transportation system applications. IET Intell Transp Syst 9(10):924–935

Fang SH, Fei YX, Xu Z, Tsao Y (2017) Learning transportation modes from smartphone sensors based on deep neural network. IEEE Sens J 16(8):6111–6118

Feng T, Timmermans H (2013) Transportation mode recognition using GPS and accelerometer data. Transp Res Part C Emerg Technol 37:118–130

Gjoreski H, Ciliberto M, Wang L, Ordonez FJ, Mekki S, Valentin S, Roggen D (2018) The university of Sussex-Huawei locomotion-transportation dataset for multimodal analytics with mobile devices. IEEE Access 6:42592–42604

Hemminki S, Nurmi P, Tarkoma S (2013) Accelerometer-based transportation mode detection on smartphones. In: Proceedings of ACM conference on embedded networked sensor systems, pp 1–14

Jahangiri A, Rakha HA (2015) Applying machine learning techniques to transportation mode recognition using mobile phone sensor data. IEEE Trans Intell Transp Syst 16(5):2406–2417

Lane ND, Miluzzo E, Lu H, Peebles D, Choudhury T, Campbell AT (2010) A survey of mobile phone sensing. IEEE Commun Mag 48(9):140–150

Ordonez FJ, Roggen D (2016) Deep convolutional and LSTM recurrent neural networks for multimodal wearable activity recognition. Sensors 16:1–25

Peng H, Long F, Ding C (2005) Feature selection based on mutual information criteria of max-dependency, max-relevance, and minredundancy. IEEE Trans Pattern Anal Mach Intell 27(8):1226–1238

Richoz S, Ciliberto M, Wang L, Birch P, Gjoreski H, Perez-Uribe A, Roggen D (2019) Human and machine recognition of transportation modes from body-worn camera images. In: Proceedings of joint 8th international conference informatics, electronics & vision and international conference on imaging, vision & pattern recognition, pp 1–6

Srivastava N, Hinton G, Krizhevsky A, Sutskever I, Salakhutdinov R (2014) Dropout: a simple way to prevent neural networks from overfitting. J Mach Learn Res 15(1):1929–1958

Su X, Caceres H, Tong H, He Q (2016) Online travel mode identification using smartphones with battery saving considerations. IEEE Trans Intell Transp Syst 17(10):2921–2934

Wang L, Gjoreski H, Ciliberto M, Mekki S, Valentin S, Roggen D (2019) Enabling reproducible research in sensor-based transportation mode recognition with the Sussex-Huawei dataset. IEEE Access 7:10870–10891

Wang L, Gjoreski H, Ciliberto M, Mekki S, Valentin S, Roggen D (2018a) Benchmarking the SHL recognition challenge with classical and deep-learning pipelines. In: Proceedings of 6th international workshop on human activity sensing corpus and applications (HASCA2018), pp 1626–1635

Wang L, Gjoreski H, Murao K, Okita T, Roggen D (2018b) Summary of the Sussex-Huawei locomotion-transportation recognition challenge. In: Proceedings of 6th international workshop on human activity sensing corpus and applications (HASCA2018), pp 1521–1530

Xia H, Qiao Y, Jian J, Chang Y (2014) Using smart phone sensors to detect transportation modes. Sensors, 20843–20865

Yu MC, Yu T, Wang SC, Lin CJ, Chang EY (2014) Big Data small footprint: the design of a low-power classifier for detecting transportation modes. In: Proceedings of very large data base endowment, pp 1429–1440

Zhang Z, Poslad S (2013) A new post correction algorithm (POCOA) for improved transportation mode recognition. In: Proceedings of IEEE international conference on systems, man, and cybernetics, pp 1512–1518

Chapter 12
Bayesian Optimization of Neural Architectures for Human Activity Recognition

Aomar Osmani and Massinissa Hamidi

Abstract Design of neural architectures is a critical aspect in deep-learning based methods. In this chapter, we explore the suitability of different neural architectures for the recognition of mobility-related human activities. Neural architecture search (NAS) is getting a lot of attention in the machine learning community and improves deep learning models' performances in many tasks like language modeling and image recognition. Deep learning techniques were successfully applied to human activity recognition (HAR). However, the design of competitive architectures remains cumbersome, time-consuming, and rely strongly on domain expertise. To address this, we propose a large-scale systematic experimental setup in order to design and evaluate neural architectures for HAR applications. Specifically, we use a Bayesian optimization (BO) procedure based on a Gaussian process surrogate model in order to tune architectures' hyper-parameters. We train and evaluate more than 600 different architectures which are then analyzed via the functional ANalysis Of VAriance (fANOVA) framework to assess hyper-parameters relevance. We experiment our approach on the Sussex-Huawei Locomotion and Transportation (SHL) dataset, a highly versatile, sensor-rich and precisely annotated dataset of human locomotion modes.

12.1 Introduction

Neural networks are attracting a considerable amount of interest in many fields achieving state-of-the-art performances. Fields like speech recognition, natural language processing, computer vision benefited largely from deep-learning techniques (He et al. 2016; Szegedy et al. 2015; Chiu et al. 2018; Luong et al. 2015; Zoph and Le 2016) and show promising results in the particular field of human activity recognition (Zeng et al. 2014; Hammerla et al. 2016; Ordóñez and Roggen 2016).

A. Osmani · M. Hamidi (✉)
Laboratoire LIPN-UMR CNRS 7030, PRES Sorbonne Paris Cité, Villetaneuse, France
e-mail: hamidi@lipn.univ-paris13.fr

A. Osmani
e-mail: ao@lipn.univ-paris13.fr

N. Kawaguchi et al. (eds.), *Human Activity Sensing*,
Springer Series in Adaptive Environments,
https://doi.org/10.1007/978-3-030-13001-5_12

One of the most important aspects behind the success of such approaches is the fact that they rely on the automatic construction of robust and highly discriminative features. In HAR specifically, many empirical results, like the work of Plötz et al. (2011), show that automatically estimated features outperform classical heuristically designed ones for many kinds of activity recognition tasks.

Robustness and discriminative power of automatically constructed features are further increased by the representational capabilities of the neural architectures. Indeed, the design of sophisticated architectures with complex connections increases the representational power of neural networks and gives the ability for the neurons to detect and learn more complex relations within, not only a given input signal but among different ones. Efforts in the particular field of computer vision are concentrated on these aspects with the development of more featured architectures. More specifically, we see the emergence of crafted designs including new kinds of connections between layers, skip connections, inception modules, convolution factorization, etc. (Szegedy et al. 2015, 2016; He et al. 2016).

However, the design of such architectures is cumbersome and, more importantly, relies heavily on domain expertise and sometimes on intuition. Difficulties to design competitive architectures stem from the high dimensionality of the neural architecture space which is impractical to explore using naive approaches. In many cases, best-performing architectures remain specialized for a given confined task, which makes it difficult to adapt to other applications.

Recently, interest has been growing in the machine learning community towards neural architecture search (NAS) which proposes various search strategies for exploring the space of neural architectures (Zoph et al. 2017; Elsken et al. 2018; Zhong et al. 2018; Pham et al. 2018). The reason for this surge of interest is explained by the competitive performances obtained by automatically designed architectures (Zoph and Le 2016; Zoph et al. 2017; Cai et al. 2018) as well as improved transfer capabilities of the architectures' building blocks between tasks (Zhong et al. 2018).

In HAR, application of NAS techniques, and deep learning approaches in general is at its infancy. Their joint potential benefits, though, are huge as pointed out by Elsken et al. (2018), who consider the application of NAS to tasks involving sensor fusion as one of its promising directions. The reason is that, in the specific case of HAR, distinguishing complex activity patterns can be performed more efficiently through the automatic learning of robust features and the automatic design of "non-intuitive" sensor fusion mechanisms.

This chapter proposes a complete large-scale setup in which we evaluate the suitability of different neural architectures for human activity recognition generated automatically using NAS techniques. Specifically, we use the conjunction of hand-crafted architectural schemes (or building blocks) and a Bayesian optimization (BO) procedure based on a Gaussian process for the surrogate model in order to adjust the neural architectures' hyper-parameters. In addition, in order to understand the influence of hyper-parameters on recognition performances and the low-level interactions among them we use the functional ANalysis Of VAriance (fANOVA) framework (Hoos and Leyton-Brown 2014). We experiment on the Sussex-Huawei Locomotion and Transportation (SHL) dataset which provides an adequate smartphone-based sensor-rich setup for the study of human locomotion modes. Part of these experiments was carried during the SHL-challenge (Gjoreski et al. 2017) and a subset of these results was published in Osmani and Hamidi (2018) as part of that challenge.

This chapter is organized as follows. Section 12.2 describes the building blocks of our architectures and how these are arranged. Description of the SHL dataset, as well as the experimental setup, are presented in Sect. 12.3. Results are presented in Sect. 12.4 which is followed by a discussion.

12.2 Proposed Approach

We propose to construct a set of different neural architectures by adjusting three main building blocks. These building blocks are related to (1) the features and the way these are extracted from an input signal using convolution layers, (2) how different data sources (modalities) are aggregated, i.e. early versus late fusion and finally, (3) how patterns and time dependencies between extracted features are detected in order to recognize specific activities.

Hyper-parameters of each of these building blocks are configured in parallel using Bayesian optimization which allows us to explore the induced neural architectures' space seeking for architectures yielding better recognition performances. In the following, we start by describing these building blocks along with the hyper-parameters being optimized before we detail the Bayesian optimization procedure and finally the assessment of the hyper-parameters importance.

12.2.1 Features Learning

The features extraction stage is based on an end-to-end stacking of various types of layers. There is no step dedicated to the manual extraction of heuristically defined features from the input signal, this one being replaced by the action of the different convolutional layers which are in some way responsible of building a hierarchy of features whose level of abstraction corresponds to the depth of the convolutional layer that was extracted from.

In the following, we consider a set of data generators or sensors each of which outputs a different modality from one another, therefore, we will use these two terms, sensors and data generators, interchangeably in the rest of this chapter. Each data generator is made-up of a given number of channels, e.g. x, y, and z axes of an accelerometer.

A convolution layer uses a set of sliding filters (or kernels) $\mathbf{w}_i \in \mathbb{R}^{ks}$ of size ks which are involved in a convolution operation with a portion of the input signal. This operation is performed on each possible portion of the input signal and outputs a set of local features. Different filters detect different features. We use a varying stride s which corresponds to the number of samples by which the filters of the convolutional layers slide over the input signals. It may be thought of as the overlap amount between two subsequent convolutions of the same signal. It is then regarded as a percentage of overlap in the hyper-parameters optimization procedure. Features produced while sliding over all possible portions of the input signal form what is called a *feature map*. This is done in parallel and yield a set of linear activations that are passed through a non-linear activation function.

The convolution operations are followed by max-pooling layers. The max-polling layers are responsible of merging previously extracted features of a given layer and selecting a subset of these according to their *level of activation*. The max pooling units have the property of being only sensitive to local maximum values and not to their exact location. This brings some good properties to the neural networks which make it invariant to different kind of small transformations of the input signal. Especially in the task we are interested in, where we care more about the fact that a given pattern is present in a sequence than its exact location (Goodfellow et al. 2016). Figure 12.2 shows a schematic representation of a complete neural architecture in which we can distinguish the features learning block encompassing a stack of convolution and max-pooling layers.

Operational constraints impose segmenting the input signals into small segments (or windows) in order to be processed by the convolutional layers. A sequence of patterns corresponding to a given locomotion mode may spread over several segments (of 1 minute in our case). In other words, for a neural network, making the correspondence between a given class and an incomplete sequence of gestures where essential parts of it are probably discarded, will make it learn something wrong which could completely mislead the learning process. We have then to make the network learns from complete sequences of patterns without potentially losing temporal dependency between different parts of a signal corresponding to the same activity.

We will assume that a given class of interest in our problem is completely defined by the sequence of patterns contained in segments of length up to 1 min and fully defined in that context.

12.2.2 Data Sources Processing

Here, we discuss how the raw sensors data are passed to the input layer of the neural network (the first convolutional layer). In other words, given the fact that modalities are composed of a number of channels, the way inputs are convoluted with the filters can vary. We can choose for example to perform the convolution on the channels taken individually or group them, etc. We define specifically three *convolutional modes* of the input sequences with each set of filters of the first layer of the base architecture (input layer):

- whole modalities grouped and convoluted, which we refer to as *grouped modalities*. See Fig. 12.1a;
- each modality convoluted apart, which is designated by *split modalities*. See Fig. 12.1b;
- each channel convoluted apart, referred to as *split channels*. See Fig. 12.1c.

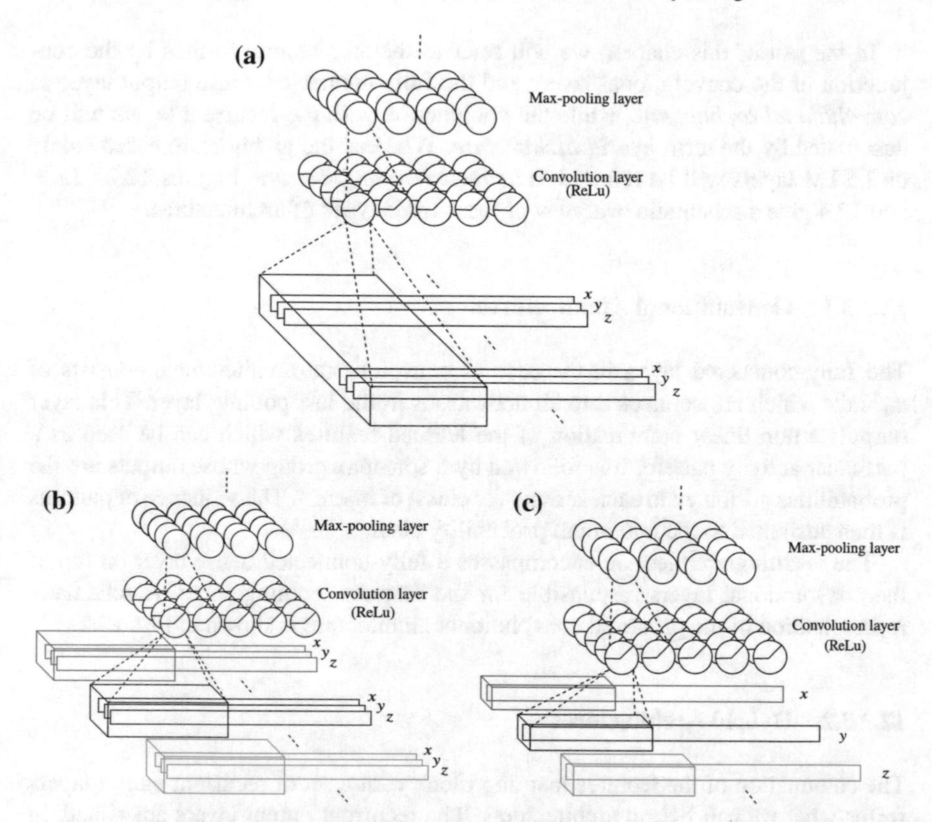

Fig. 12.1 Schematic representation of the different *convolutional modes* of input data. **a** Modalities are grouped together and convoluted with the filters. **b** Modalities are taken apart from each other. **c** Each channel is convoluted alone

Note, in Fig. 12.1, that depending on how the convolution is performed, a concatenation procedure of the learned features vectors will be required before getting to the next layers. This is, for example, the case for the split channels convolutional mode. This way, the features are extracted at different levels depending on the convolutional mode. In other words, the max-pooling layers, which are responsible for selecting the units with maximum activation values through a given sequence, will process each modality or channel separately, and the resulting characteristics will necessarily be different from one convolutional mode to another.

12.2.3 Classification

Specific activity patterns within, either previously extracted features, or within the raw input signals, are detected using two types of layers: (1) a fully-connected dense layer and (2) recurrent long short-term memory (LSTM) layers (Hochreiter and Schmidhuber 1997).

In the rest of this chapter, we will refer to the architecture formed by the conjunction of the convolutional layers and the fully-connected dense output layer as *convolutional architecture*, while the conjunction with the recurrent layers will be designated by the term *hybrid architecture*. Whereas the architecture based solely on LSTM layers will be referred to as *recurrent architecture*. Figures 12.2, 12.3, and 12.4 give a schematic overview of these three types of architectures.

12.2.3.1 Convolutional Architectures

The fully-connected layer, in the case of convolutional architectures, consists of n_u units which are connected to all activations in the last pooling layer. This layer outputs a non-linear combination of the learned features which can be seen as a particular activity pattern. It is followed by a soft-max group whose outputs are the probabilities attributed to each activity (or class) of interest. The sequence of patterns is then attributed to the maximum probability class.

The resulting architecture encompasses a fully-connected dense layer on top of the convolutional layers responsible for the features learning stage. A schematic representation of the proposed convolutional architecture is shown in Fig. 12.2.

12.2.3.2 Hybrid Architectures

The conjunction of the features learning block with a set of recurrent output layers forms what we call hybrid architectures. The recurrent output layers are aimed, in our case, at capturing the time dependency that would exist in the input sequences and in particular, between the vectors of characteristics learned previously through convolutional layers. This is even more important when the size of the sequences is very large, heterogeneous and when some patterns appear at several points in the sequence. Due to their sequence-based specialization, the recurrent layers tend to scale-up easily unlike the convolutional ones that remain somehow shallow. In particular, one of the types of recurrent cells that present this ability to scale-up is the Long Short-Term Memory (LSTM) unit.

LSTM units (Hochreiter and Schmidhuber 1997) are a powerful type of recurrent neural networks that circumvent the long-term dependency problem when it comes to memorizing pieces of information through long periods. Indeed, one of the common problems of recurrent networks arises from the exponentially smaller weights given to long-term interactions compared to short ones when dealing with long sequences. In order to circumvent to these problems, the LSTM units introduce a set of mechanisms and gates to produce paths where the gradient can flow for long duration (Goodfellow et al. 2016).

The main component of these units is the cell state that is designed in the goal of retaining information through long periods of time. Information is added to and removed from this cell state using different gates. There are four gates in total that interact and decide at each time-step which information is carried over from the

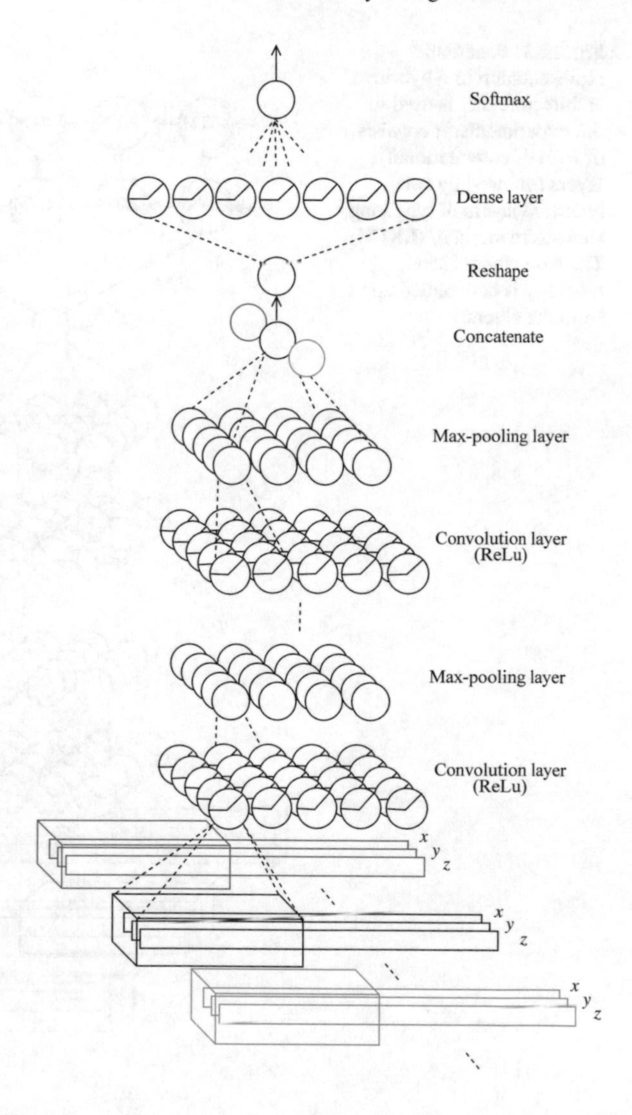

Fig. 12.2 Schematic representation of a convolutional neural network that is used in our experiments. It encompasses a set of convolutional layers followed by a dense layer. The case where each modality is convoluted apart from the others. The features maps outputted from the last convolutional layer of each modality is concatenated and then reshaped in order to be fed into the dense layer

previous time-step, which components of the current input should be kept, and finally which parts of the cell state should be outputted to the next time-step.

The resulting architecture, shown in Fig. 12.3, bears a close resemblance to that proposed by Yao et al. (2017) in which the author experimented a similar construction for different applications such as car tracking with motion sensors, and user identification with motion analysis. However, their experiments were confined solely to data from accelerometers and gyroscopes.

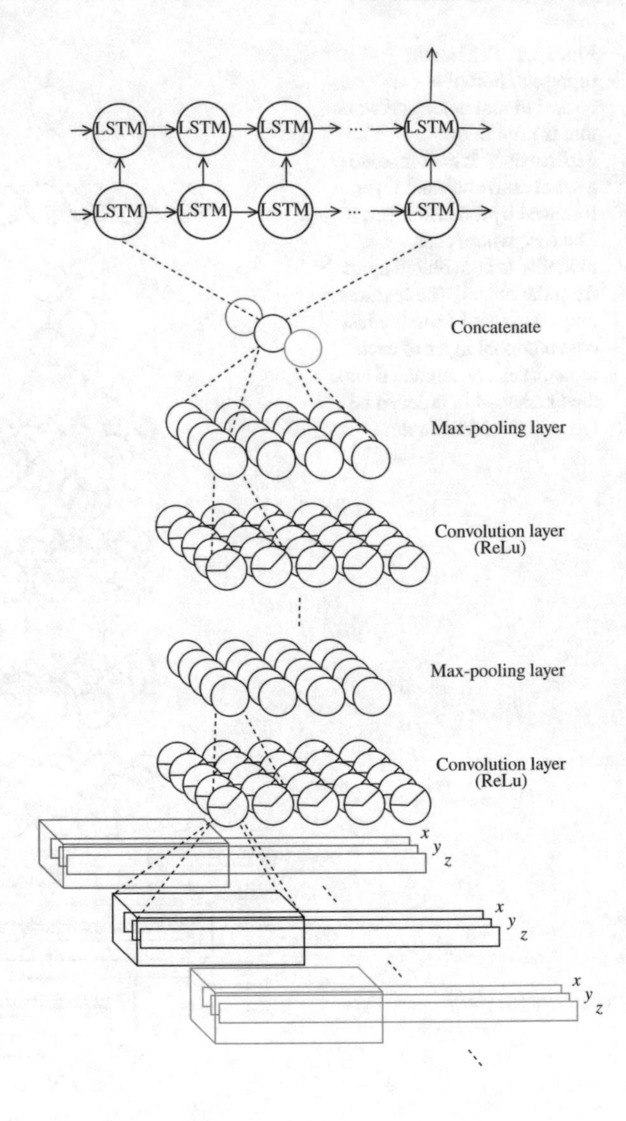

Fig. 12.3 Schematic representation of a hybrid architecture that is used in our experiments. It consists of a set of convolutional layers followed by two recurrent layers of type long short-term memory (LSTM). The case where each modality is convoluted apart from the others

12.2.3.3 Recurrent Architectures

The third architecture that we propose is based on recurrent (LSTM) units solely. In contrasts with hybrid architectures, these units process the input signals as they are without any additional features extraction step.

We opted for this kind of architectures because of the sequential nature of inputs and the suitability of such units to learn temporal dependencies of the inputs. Indeed, besides the additional features learning stage, this architecture contrasts with the hybrid one in two ways; the first one is related to the kind of patterns that are learned. Here, raw variations of the signals are taken into account, while in the

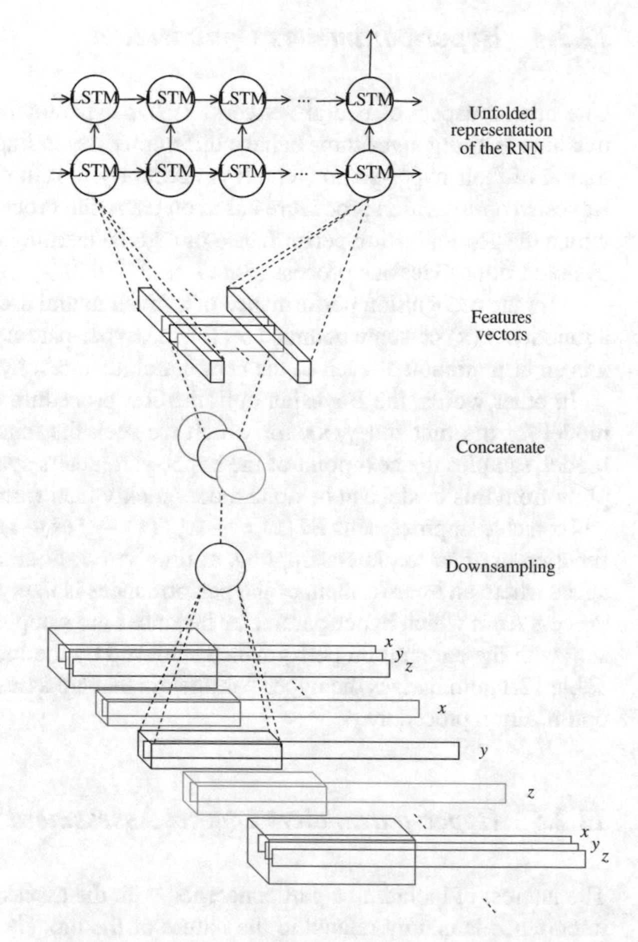

Fig. 12.4 Schematic representation of a recurrent architecture that is used in our experiments. Unlike convolutional and hybrid architectures, this architecture does not encompass an explicit feature extraction step this one being performed through the hidden states of the LSTM units that are updated across the input signal

hybrid architectures, variations of high-level features are considered. The second difference is related to the size of the patterns that are learned. Indeed, here, the sequence is much longer than in the case of hybrid architectures. This makes the task of learning patterns for the recurrent layers more difficult.

From a technical perspective, inputs are first down-sampled by a given factor. Down-sampling was chosen because it is one of the most practical ways to fit with the computational requirements that are imposed when using recurrent models. This way, we alleviate the recourse to any additional segmentation process and comply with the assumption made above about input lengths. Each channel is processed at a time before being concatenated and fed into the LSTM units.In this case, it is the joint variations of the different modalities that are being learned by the recurrent layers. Figure 12.4 shows a schematic representation of the resulting architecture.

12.2.4 Hyper-parameters Optimization

One crucial aspect of neural networks is hyper-parameters tuning. Indeed, as the machine learning algorithms behave differently depending on the particular instantiation of their hyper-parameters, it is necessary to refine them. For this, we use a Bayesian optimization procedure based on Gaussian process as a surrogate model in which the generalization performance of a given learning algorithm is modeled as a sample from a Gaussian process (Snoek et al. 2012).

Here the recognition performance of a given neural architecture is considered as a function $f(\mathbf{x})$ on some bounded set $\mathcal{X}$ (the hyper-parameters search space), where $\mathbf{x}$ is an instantiation of each of the neural architecture's hyper-parameters.

In other words, the Bayesian optimization procedure constructs a probabilistic model for the function $f(\mathbf{x})$, for which we seek the minimum, and based on that model, samples the next point of the hyper-parameters space where to search. Sampling from this model can be done more quickly than from the original function.

Expected improvement $EI(\mathbf{x}) = -\mathbb{E}[f(\mathbf{x}) - f(\mathbf{x} + t)]$ is used as an acquisition function in order to direct sampling, at time step t, of areas of the hyper-parameter space where an improvement of the performances is likely to happen. The Gaussian process from which hyper-parameter instances are sampled is updated at each time step with the recognition performance achieved by the induced neural architecture. Table 12.1 summarizes the hyper-parameters that are assessed through the Bayesian optimization procedure.

12.2.5 Hyper-parameters Impact Assessment

The interest of including a part concerned with the assessment of hyper-parameters importance is mainly related to the nature of the models that we propose to study. Deep neural networks are the typical example of so-called *black-box* functions. The work consisting of interpreting the behavior and results of these models is an important aspect.

In particular, we want to investigate more closely and analyze the link between the different parameters of configurations such as the link between the number of cells in the different LSTM layers or the relations between the filters sizes of the different convolutional layers.

We use for this purpose the functional ANalysis Of VAriance (fANOVA) framework (Hoos and Leyton-Brown 2014) dedicated to the diagnosis of functions of dependent variables in high-dimensional spaces. These functions include, in particular, the context of machine learning whether for exploration and analysis of predictors of a dataset or in our case, the analysis of the relative importance of hyper-parameters with regards to the architecture's performances and their interactions. This framework has been used by Hammerla et al. (2016) in the context of activity recognition and works by, first, fitting a random forest model on the obtained results from the BO procedure then, constructing a function of the specified hyper-parameters of interest and marginalizing the effect of the others.

12.3 Experiments

In order to evaluate the proposed approach, we conduct experiments on a subset of the Sussex-Huawei Locomotion-Transportation dataset. Code to reproduce the experiments is publicly made available.[1]

We use the computing resources of a dedicated cluster called *Magi*. This grid is composed of several nodes of 40 cores each with an execution speed of 2.30 GHz and a working memory of 64 GB. In order to determine the best combination, we let a given configuration of the BO procedure run for 30 iterations in total.

Models development and BO procedures are based on off-the-shelf implementations, all of which are free software. In particular, we use the Tensorflow framework (Abadi et al. 2016) for building the deep neural networks but also the `scikit-learn` library (Pedregosa et al. 2011) and the `scikit-optimize` library specialized in optimizing cost functions. The implementation of the fANOVA framework proposed by Hoos and Leyton-Brown (2014) is also free software and publicly available.

12.3.1 SHL Dataset and Task Description

The SHL dataset (Gjoreski et al. 2018)[2] is a highly versatile and precisely annotated dataset aiming to overcome the lack of such datasets dedicated to transportation (750 h of labeled locomotion data). The SHL dataset contains multi-modal locomotion data recorded in real-life settings. There are in total 16 modalities including accelerometer, gyroscope, cellular networks, WiFi networks, audio, etc. making it suitable for a wide range of applications and in particular the task we are interested in which is transportation recognition. Indeed, there are 8 primary categories of transportation that we are interested in: Still, Walk, Run, Bike, Car, Bus, Train, Subway (Tube).

In our experiments, we restrict our study to a subset of the SHL dataset. This subset corresponds to data recorded by a single participant (user 1) using a single smartphone carried inside the front right pocket. The participant was performing the activities on a daily basis (approximately 5–8 h per day). Sensor signals were sampled at 100 Hz and the frames, for both training and testing dataset were generated by segmenting the whole data with a non-overlap sliding window of 1-min length. The subset that we use consists of 16,310 sequences, which corresponds to a total of almost 5 h for the training part. Next to that, 5698 sequences of a duration of 1 min each are reserved to test our models. This subset corresponds to 82 days of data collection; 20 days for testing and 62 days for training.

Among the 16 modalities of the original dataset, 7 are used in our experiments: accelerometer, gyroscope, magnetometer, linear acceleration, orientation, gravity, and ambient pressure.

---

[1] https://www.github.com/hamidimassinissa/hasca-shl.

[2] The preview of the SHL data set can be downloaded from: http://www.shl-dataset.org/download/.

12.3.2 Experimental Setup

Here we detail precisely the different building blocks presented in Sect. 12.2. Note that in the following, hyper-parameters accompanied with a mathematical notation are subject to the BO procedure. We use up to 3 convolutional layers in the features learning stage, each followed by a unit of max-pooling whose parameters, i.e. window size, is set to 2. Each convolutional layer has its own set of filters that have sizes ks_i, $i \in \{1, 2, 3\}$ and are different from one layer to another. The number of filters n_f meanwhile remains the same for each layer of a given architecture. We use two types of activation functions, namely the rectified linear unit (ReLU) and the hyperbolic tangent (Tanh) activation functions. Concerning the recurrent layers, we use 2 layers of LSTM units composed of n_{hu1} and n_{hu2} hidden units for layers 1 and 2 respectively. Moreover, we initialize the bias of the forget gate to 1 according to Jozefowicz et al. (2015) who recommend setting the bias of this gate to relatively wide values such as 1 or 2 which allows the gradient to flow easily.

With regard to the input signals, according to the assumption we made in Sect. 12.2, we decided to discard any additional segmentation process that could potentially introduce bias into the learning process. Inputs are, then, taken as they are without any additional segmentation process. The entries of our different architectures are therefore of the order of 1 min, i.e. 6000 samples, given that the sampling rate is 100 Hz.

12.3.3 Training Details

We use batch normalization on top of each convolutional layer and apply it on each input regardless of the convolutional mode that is used. This procedure makes the neural networks more stable by normalizing the inputs data in each batch (Ioffe and Szegedy 2015). Note that this standardization procedure is more than necessary. Training the models without this procedure results in poor recognition performances.

In order to prevent our architectures from overfitting the training data, we use dropout. We apply specifically in the case of the LSTM units, dropout according to the *variational RNN* technique presented in Gal and Ghahramani (2016). In contrast with the naive regularization method, this technique applies the same mask not only to the input and output connections but also to the recurrent connections. Three parameters in total, which are assessed by the way in the BO procedure, are used for this technique namely, input p_{in}, output p_{ou}, and state p_{st}, keep probabilities. Dropout p_d is also applied on top of the fully-connected layer in the case of convolutional architectures and is also assessed through the BO procedure.

The learning rate lr, which controls how coefficients are updated at each step of the training phase is assessed. The learning decay is fixed to 0.1 and controls the amount by which the learning rate is decreased along epochs.

12.4 Results

Starting from the three types of architectures defined in Sect. 12.2 and the BO settings shown in Table 12.1, we end up with no less than 600 different architectures. These architectures are trained for 13 successive epochs on the training data and evaluated on the validation set after each epoch. The training phase is stopped after 3 subsequent epochs without any improvement of the recognition performances. We use hold-out for model evaluation and the f1-score in order to assess architectures' recognition performances. The f1-score is computed as follows:

$$F1 = 2 \times \frac{Re \cdot Pr}{Re + Pr}$$
$$Pr := \frac{TP}{TP + FP}$$
$$Re := \frac{TP}{TP + FN}$$

where TP, FP, and FN are the true positives, false positives, and false negatives rates respectively. This metric has the advantage of giving more realistic insights about the architectures' performances but presents some weaknesses when it is confronted with strongly unbalanced datasets which introduce bias depending on the way this metric is calculated, i.e. averaging f1-score in each fold, averaging over precision and recall in each fold, etc. (Forman and Scholz 2010). Given the absence of any noticeable imbalance in our dataset, we do not dwell on this topic anymore in the following, but we are keen to point this aspect out as it is often disregarded.

Table 12.1 Summary of the different hyper-parameters assessed during Bayesian optimization procedure along with their respective bounds

Hyper-param. (sym)	Low	High	Prior
Learning rate (lr)	0.001	0.1	log
Kernel size 1st (ks_1)	9	15	–
Kernel size 2nd (ks_2)	9	15	–
Kernel size 3rd (ks_3)	9	12	–
Number of filters (nf)	16	28	–
Stride (s)	0.5	0.6	log
Dropout probability (p_d)	0.1	0.5	log
Number of units dense layer (n_u)	64	2048	–
Number of hidden units 1 (n_{hu1})	64	384	–
Number of hidden units 2 (n_{hu2})	64	384	–
Inputs dropout probability (p_{in})	0.5	1	log
Outputs dropout probability (p_{ou})	0.5	1	log
States dropout probability (p_{st})	0.5	1	log

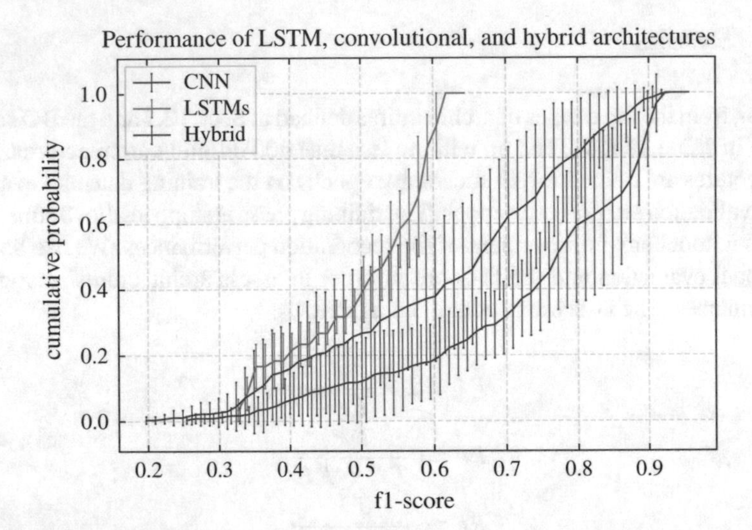

Fig. 12.5 Cumulative distribution of the recognition performances for hybrid, convolutional, and recurrent architectures. Recognition performances are presented in the form of cumulative distributions obtained through the whole runs of the Bayesian optimization procedures

12.4.1 Comparison of Architectures' Performances

Recognition performances of the different configurations are presented in the form of cumulative distributions. These graphs are intended to provide a sense of the underlying hyper-parameter spaces induced each time by the various architectures.

Figure 12.5 shows a comparison between the recognition performances of each type of architecture obtained through the whole runs of the BO procedures. Results reveal that hybrid architectures outperform slightly the convolutional architectures and by far the recurrent ones. Best architectures, both hybrid and convolutional, achieve more than 91% f1-score and corroborate results of approaches that rely on features learning for recognizing human activities, e.g. Zeng et al. (2014), Hammerla et al. (2016), Ordóñez and Roggen (2016).

In the other hand, best recurrent architectures achieve an f1-score of 64% and overall, these architectures perform poorly in recognizing human activities. We suspect that this is related to the constraint imposed on the length of the inputs which probably causes the hidden states to not flow easily through time. Investigating these aspects requires further experiments. Surprisingly, although hybrid and convolutional architectures achieve the best recognition performances, these exhibit a large spread in terms of performances which range from 0.2 to 0.92, suggesting that the induced hyper-parameter space is characterized by a certain complexity contrasting with the relatively constrained architectures' configurations that we set up. This is further confirmed in the particular case of convolutional architectures which reveal a consequent variance in terms of performances that is worth ± 0.4 all over the performances range.

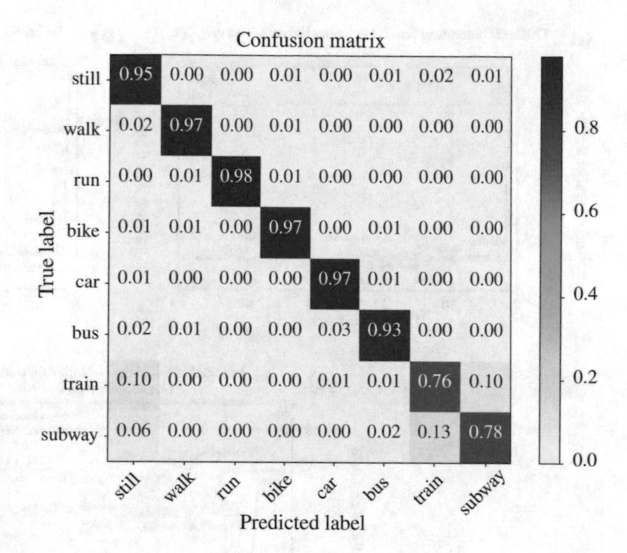

Fig. 12.6 Confusion matrix for the 8 activities obtained with a hybrid model trained on all modalities using a tenfold cross-validation procedure. The architecture relies on the split channels convolutional mode and apply dropout to the recurrent layers as well as batch normalization to the inputs. Hyper-parameters of the best performing architecture are: $n_{hu1} = 176$, $n_{hu2} = 384$, $lr = 0.1$, $nf = 20$, $ks_i \in \{10, 13, 10\}$, $s = 0.558$

Figure 12.6 shows the recognition performances of the best performing architecture in the form of a confusion matrix. This time, the architecture was evaluated with a tenfold cross-validation procedure on the entire dataset. It should be noted from the figure that there is an overlap between the *train* and *subway* classes which reflects a clear link between the two activities. Apart from these two classes, the model achieves good recognition rates calculated via the f1-score and which are in the order of 90% overall.

12.4.2 Convergence of the Bayesian Optimization Procedure

Convergence plots of the Bayesian optimization procedure for different configurations are presented in Fig. 12.7. Even though these procedures are characterized by non-deterministic aspects, which are by the way studied by Wang and de Freitas (2014), the study of their spatial and temporal behavior can give insights about the underlying neural architectures space.

Overall, the BO procedure performs well in various configurations and converges toward the optimum in only a few iterations (Fig. 12.7b). This is the case for example for split channels and split modalities where a region of optimal recognition performances is obtained after four runs of the BO procedure while making a sustainable improvement. This contrasts with all modalities grouped in which case the BO procedure is trapped in a large plateau of equivalent architectures. Curiously, in the case where subsets of modalities are used, the BO procedure converges gradually after getting stuck in many successive plateaux. Figure 12.7c shows the convergence of the BO procedure in the case of hybrid architectures trained on different modalities subsets.

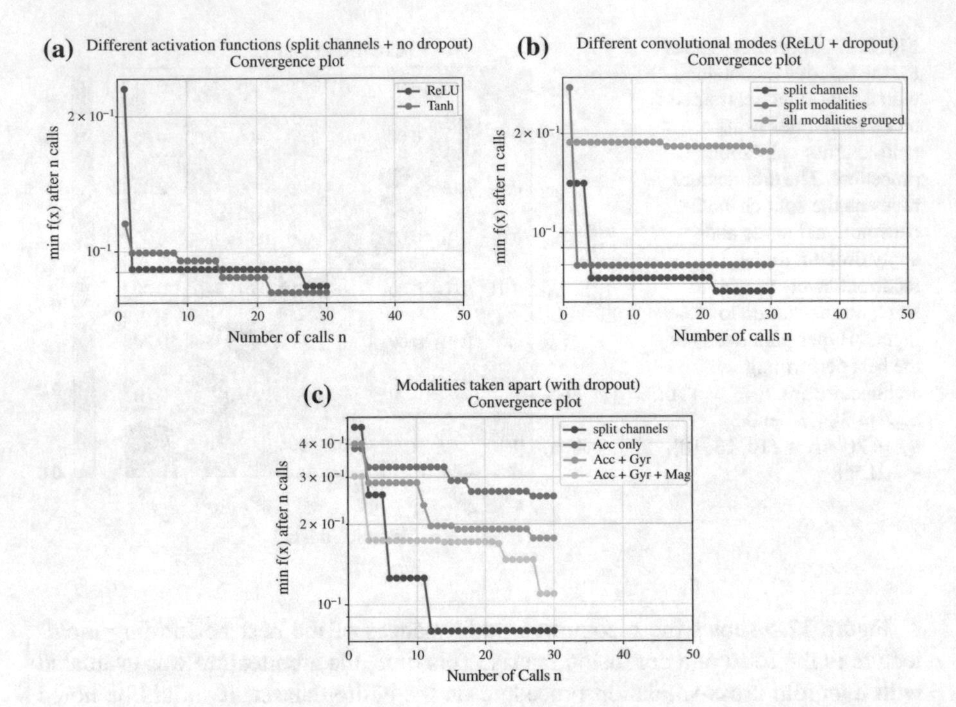

Fig. 12.7 Convergence plots of the Bayesian optimization procedure in various configurations showing specifically the influence of: **a** the activation functions (ReLU vs. Tanh), **b** the convolutional mode in the case of convolutional architectures, and **c** the conjunction of different subsets of modalities in the case of hybrid architectures. Y-axis grows on a logarithmic scale

12.4.3 Performances of the Features Learning Stage

Features learning is a critical part of the convolutional and hybrid architectures. In our case, learned features and, to a greater extent, recognition performances are not solely influenced by the convolutional mode being used. Rather, many interactions do account during the features learning stage involving in one hand the activation function and in the other hand the stride, the kernel sizes, the number of filters, the number of convolutional layers, etc.

Figure 12.8 compares the influence of the type of activation functions being used in the features learning stage on the cumulative distributions of convolutional architectures. Overall, we notice that the type of activation function doesn't have a significant impact on the recognition performances of both convolutional and hybrid architectures. However, we do notice in the case of split channels that recognition performances of convolutional architectures with a ReLU function have a larger spread (0.25–0.9) than the ones with a Tanh function (0.5–0.9).

This contrasts with all modalities grouped and split modalities in which recognition performances spread equivalently regardless of the type of activation function being used. Interestingly, additional experiments using only subsets of the previ-

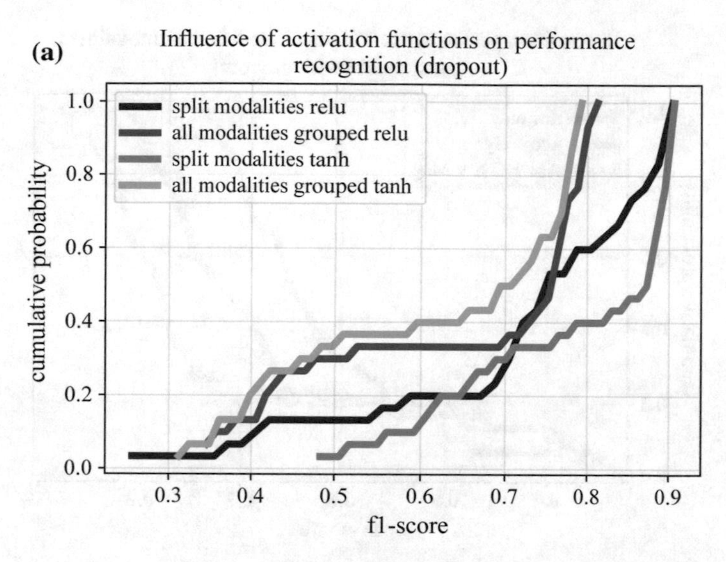

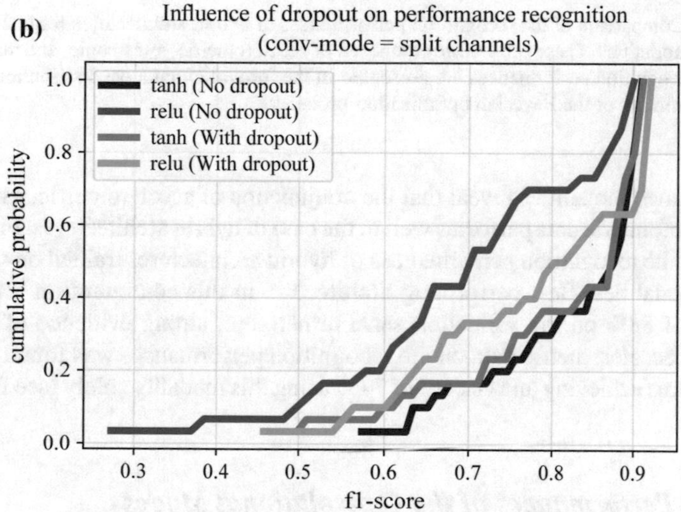

Fig. 12.8 Cumulative distribution of recognition performances of convolutional architectures in multiple configurations showing, in particular, the influence of the type of activation function being used. **a** Depicts split modalities and all modalities grouped convolutional modes together and **b** split channels along with the application of dropout or not on the dense layer. Recognition performances are presented in the form of cumulative distributions obtained during a given run of the Bayesian optimization procedure

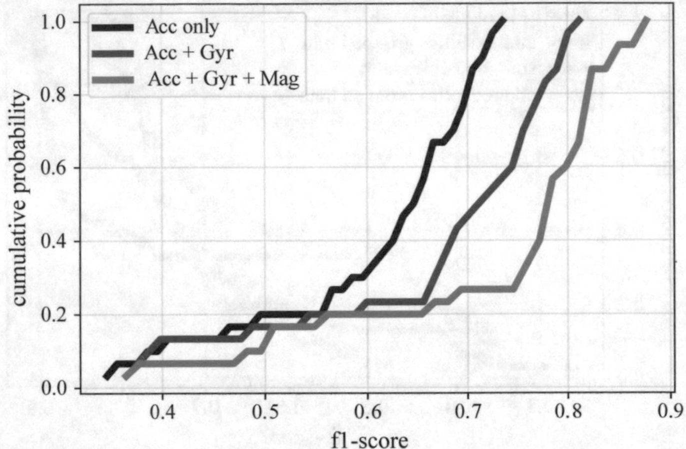

Fig. 12.9 Comparison of the recognition performances of hybrid architectures trained on various subsets of modalities. These subsets are composed of accelerometric, gyroscopic, and magnetometric data. Recognition performances are presented in the form of cumulative distributions obtained during a given run of the Bayesian optimization procedure

ously retained modalities reveal that the conjunction of accelerometric, gyroscopic, and magnetometric data performs well in the case of hybrid architectures. Figure 12.9 compares the recognition performances of hybrid architectures trained on small subsets of modalities. Best performing architecture in this configuration achieves an f1-score of 88% on the validation set. Furthermore, strong evidence of the huge impact of accelerometric data on the recognition performances was found with best architectures achieving an f1-score of 74% using this modality solely (see Fig. 12.9).

12.4.4 Performances of the Convolutional Modes

Figures 12.10 and 12.11 show recognition performances achieved by hybrid and convolutional architectures, respectively. These figures compare in particular the influence of the different convolutional modes being experimented in this work. In the other hand, Figs. 12.12 and 12.13 provide pairwise marginal plots of a set of hyper-parameters in the case of convolutional and hybrid architectures respectively which are obtained through the fANOVA framework.

Taken as a whole, recognition performances, characterized by the cumulative distributions shown in Figs. 12.10 and 12.11, exhibit a slightly similar trend and this, regardless of the convolutional mode. More specifically, in the case of architectures with split modalities, throughout the BO runs we do notice that the number of archi-

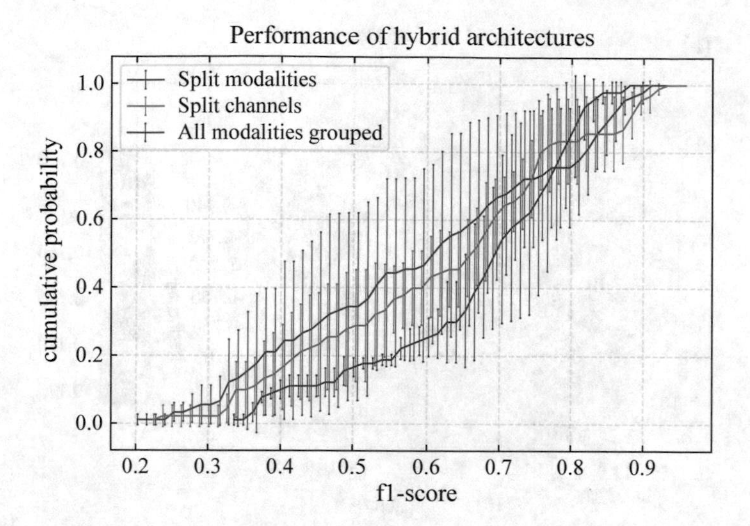

Fig. 12.10 Cumulative distribution of recognition performances for hybrid architectures showing the impact of each convolutional mode. Recognition performances of hybrid architectures averaged throughout the whole runs of the Bayesian optimization procedures and grouped based on the corresponding convolutional mode

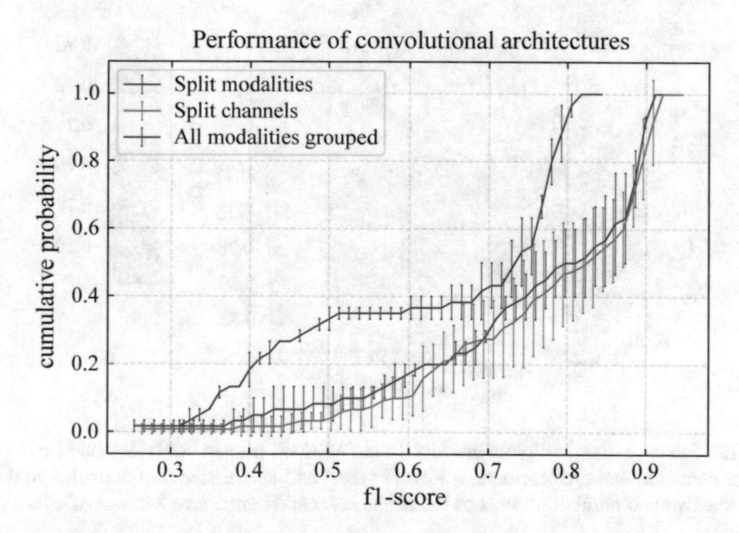

Fig. 12.11 Cumulative distribution of recognition performances for convolutional and hybrid architectures along with the different convolutional modes. Recognition performances of convolutional architectures averaged throughout the whole runs of the Bayesian optimization procedures and grouped based on the corresponding convolutional mode

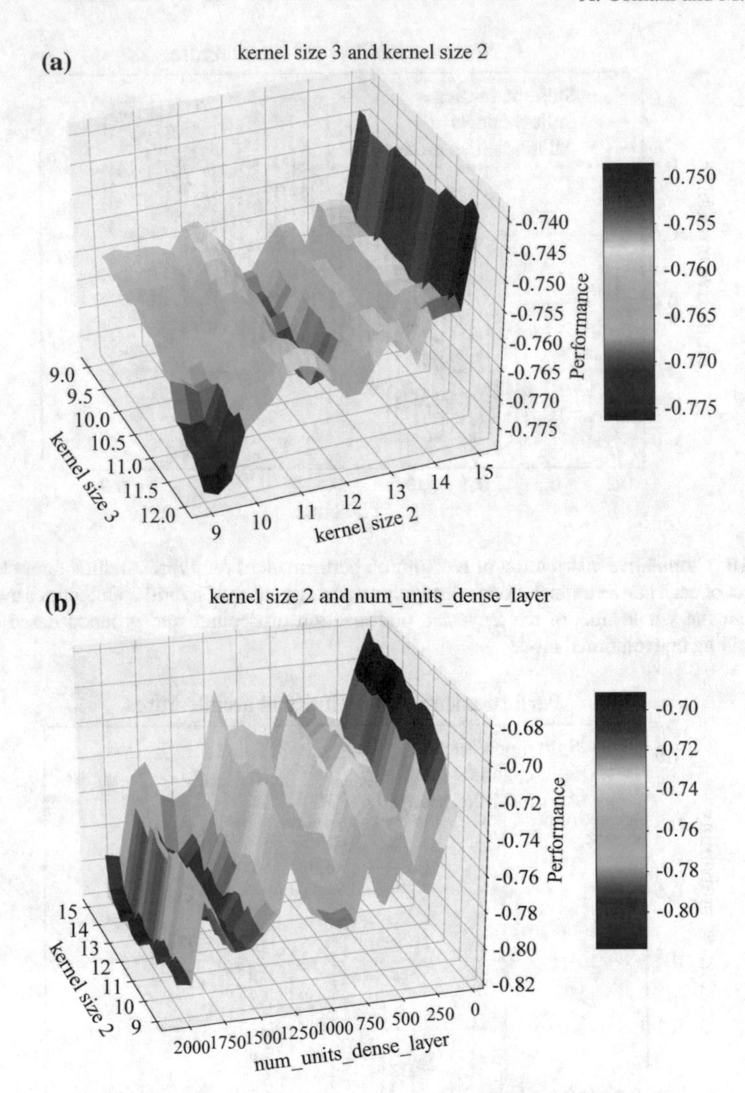

Fig. 12.12 Pairwise marginal plots produced via fANOVA framework (Hoos and Leyton-Brown 2014) for convolutional architectures. **a** Kernel size 2 and kernel size 3 of convolutional layers 2 and 3 respectively, **b** number of units of the dense layer and kernel size 2 of convolutional layer 2

tectures yielding, for example, a given recognition performance varies to a critical large extent (± 0.6 standard deviation points). This underlines just how complex the subspace generated by the split modalities convolutional mode is. In other words, the generated subspace can be viewed as spanning the entire recognition performances range and encompassing many plateaus of variable sizes scattered all over it, each of which yielding equivalent performing architectures.

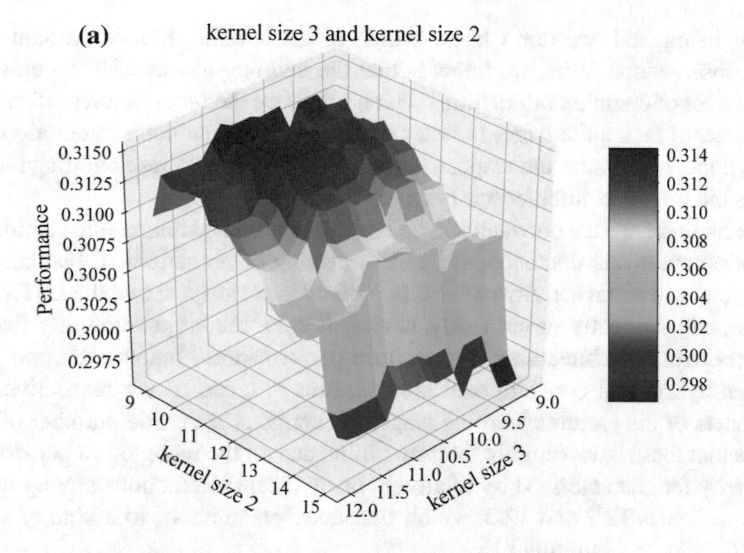

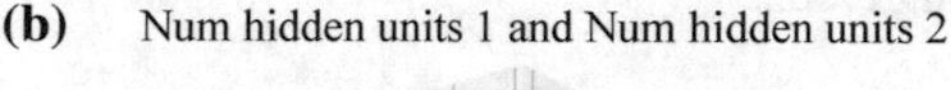

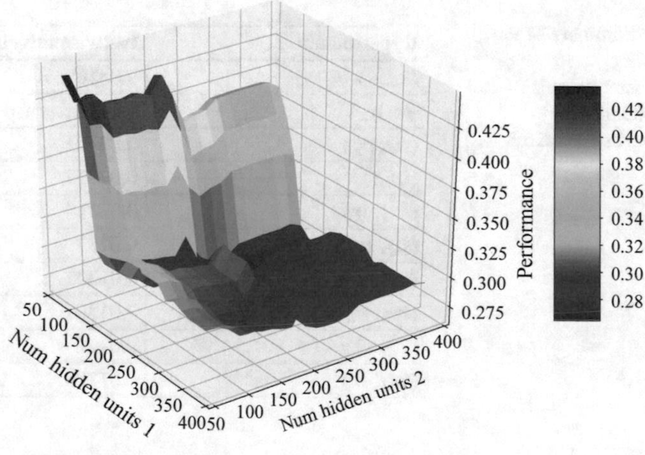

Fig. 12.13 Pairwise marginal plots produced via fANOVA framework (Hoos and Leyton-Brown 2014) for hybrid architectures. **a** Kernel size 2 and kernel size 3 of convolutional layers 2 and 3 respectively, **b** number of hidden units in LSTM layers 1 and 2 respectively

This contrasts with the convolutional architectures where we do not get a similar variability (Fig. 12.11). Rather, we see clearly that, with the exception of all modalities grouped convolutional mode which shows a much more stable behavior, the variability is confined to ±0.2 standard deviation points. This suggests that the entire hyper-parameter space induced by convolutional architectures, regardless of the convolutional mode, is characterized by abrupt topology and the absence of sustainable plateaus which is presumably what the analysis of the pairwise marginal plots shown in Fig. 12.12 tends to support to a certain extent.

This being said, we don't know which of the building blocks account for the above observations. What we know is that the building blocks of our architectures exhibit a lot of complex interactions which influence the recognition performances, as a matter of fact, quite large. In order to get insights about these interactions, what can be done, is to assess the respective hyper-parameters of these building blocks as well as the low-level interactions between them.

The main difference between hybrid and convolutional architectures reside in the type of output layers that accounts for discovering combinations of features which correspond to a given activity pattern. In particular, we suspect that the LSTM layers may be, either directly or indirectly, responsible for the large variability that characterizes hybrid architectures' recognition performances. Indeed, the comparison between hybrid and convolutional architectures in terms of the respective hyper-parameters of the features learning stage, i.e. kernel sizes, stride, number of filters, reveals that these have roughly the same influence on the recognition performances and are by far characterized by relatively insignificant interactions among them, as shown in Tables 12.2 and 12.3, which translates, presumably, to a kind of stability characterizing this building block.

Table 12.2 Summary of the most important hyper-parameter pairwise marginals obtained on convolutional architectures

Hyper-param.	Pairwise marginal
(n_u, s)	0.0516
(lr, n_u)	0.0478
(p_d, n_u)	0.02601
(lr, s)	0.02577
(ks_1, lr)	0.02114
(p_d, lr)	0.0181
(n_f, n_u)	0.01341
(ks_3, n_u)	0.01301
(p_d, s)	0.0122
(ks_2, lr)	0.01209

Table 12.3 Summary of the most important hyper-parameter pairwise marginals obtained on hybrid architectures

Hyper-param.	Pairwise marginal
(n_{hu1}, lr)	0.0326
(lr, s)	0.0298
(n_{hu2}, s)	0.0292
(n_{hu2}, lr)	0.0267
(n_{hu1}, s)	0.0239
(n_{hu1}, n_{hu2})	0.02208
(lr, n_f)	0.0172
(ks_3, lr)	0.015001
(n_{hu2}, ks_3)	0.01413
(ks_2, lr)	0.01354

Table 12.4 Summary of the hyper-parameters' importance obtained through the fANOVA analysis of convolutional and hybrid architectures' recognition performances

Hyper-param. (sym)	Individual importance	
	Convolutional	Hybrid
Learning rate (l_r)	0.10423	0.19815
Kernel size 1st (ks_1)	0.01410	0.00874
Kernel size 2nd (ks_2)	0.00916	0.023105
Kernel size 3rd (ks_3)	0.04373	0.01788
Number of filters (n_f)	0.02810	0.01845
Stride (s)	0.08092	0.06236
Dropout probability (p_d)	0.03279	–
Number of units dense layer (n_u)	0.16748	–
Number of hidden units 1 (n_{hu1})	–	0.06324
Number of hidden units 2 (n_{hu2})	–	0.02478
Inputs dropout probability (p_{in})	–	0.04047
Outputs dropout probability (p_{ou})	–	0.01056
States dropout probability (p_{st})	–	0.01991

In the other hand, dense layers' hyper-parameters exhibit an extremely different impact. Indeed, the number of units in the dense layer n_u is found to be the most important hyper-parameter in the case of convolutional architectures and accounts for more than 16% of the recognition performances variability. In addition, the number of units in the dense layer has the largest low-level interactions (see Table 12.2), which confirms the important impact of the dense layer in recognizing activity patterns. This contrasts with the quantified importance of the number of hidden units of LSTM layers which is 6% and 2% for the first and second layer respectively. We notice instead that the learning rate is taking advantage over the number of hidden units and accounts for more than 19% of the recognition performances variability in the case of hybrid architectures (Table 12.4).

Although this has to be taken with care and need to be further investigated as it involves an optimization procedure which, we should recall, is characterized by non-deterministic behavior, our findings would seem to imply that the dense layers exhibit more representational power compared to LSTM layers, the latter being disadvantaged during the BO procedure in favor of the learning rate. This observation corroborates the findings of Greff et al. (2017) and Hammerla et al. (2016) in both a general learning setting and a HAR-specific one respectively.

12.5 Discussion

We explored in this chapter the suitability of different neural architectures for the recognition of mobility-related human activities. Precisely, we experimented with three different types of neural architectures namely, convolutional, hybrid and recurrent. Various configurations of these architectures are generated through the combination of different convolutional modes and exploration of the hyper-parameters space using neural architecture search techniques. These configurations are then analyzed with the fANOVA framework.

Overall, we obtained good recognition performances with a certain advantage for convolutional and hybrid architectures over the recurrent ones. The joint exploitation of deep learning approaches and neural architecture search techniques is by far beneficial for HAR applications. Surprisingly, even though we restricted the neural architecture space with a limited number of architectural schemes, the exploration of the induced spaces reveals a substantial variability in terms of performances. This makes the case for further exploration of neural architectures spaces characterized by more degrees of freedom and more architectural schemes in order to get a deeper sense on transfer capabilities and sensor fusion aspects of these approaches.

References

Abadi M et al (2016) Tensorflow: a system for large-scale machine learning. OSDI 16:265–283

Cai H, Chen T, Zhang W, Yu Y, Wang J (2018) Efficient architecture search by network transformation. AAAI

Chiu C-C, Sainath TN, Wu Y, Prabhavalkar R, Nguyen P, Chen Z, Kannan A, Weiss RJ, Rao K, Gonina E et al (2018) State-of-the-art speech recognition with sequence-to-sequence models. In: 2018 IEEE international conference on acoustics, speech and signal processing (ICASSP). IEEE, pp 4774–4778

Elsken T, Metzen JH, Hutter F (2018) Neural architecture search: a survey. arXiv:1808.05377

Forman G, Scholz M (2010) Apples-to-apples in cross-validation studies: pitfalls in classifier performance measurement. ACM SIGKDD Explor Newsl 12(1):49–57

Gal Y, Ghahramani Z (2016) A theoretically grounded application of dropout in recurrent neural networks. In: Advances in neural information processing systems, pp 1019–1027

Gjoreski H, Ciliberto M, Morales FJO, Roggen D, Mekki S, Valentin S (2017) A versatile annotated dataset for multimodal locomotion analytics with mobile devices. In: Proceedings of the 15th ACM conference on embedded network sensor systems. ACM, p 61

Gjoreski H, Ciliberto M, Wang L, Morales FJO, Mekki S, Valentin S, Roggen D (2018) The University of Sussex-Huawei locomotion and transportation dataset for multimodal analytics with mobile devices. IEEE Access

Goodfellow I, Bengio Y, Courville A, Bengio Y (2016) Deep learning, vol 1. MIT Press, Cambridge

Greff K, Srivastava RK, Koutník J, Steunebrink BR, Schmidhuber J (2017) LSTM: a search space odyssey. IEEE Trans Neural Netw Learn Syst 28(10):2222–2232

Hammerla NY, Halloran S, Plötz T (2016) Deep, convolutional, and recurrent models for human activity recognition using wearables. In: International joint conference on artificial intelligence, pp 1533–1540

He K, Zhang X, Ren S, Sun J (2016) Deep residual learning for image recognition. In: Proceedings of the IEEE conference on computer vision and pattern recognition, pp 770–778

Hochreiter S, Schmidhuber J (1997) Long short-term memory. Neural Comput 9(8):1735–1780

Hoos H, Leyton-Brown K (2014) An efficient approach for assessing hyperparameter importance. In: International conference on machine learning, pp 754–762

Ioffe S, Szegedy C (2015) Batch normalization: accelerating deep network training by reducing internal covariate shift. In: International conference on machine learning, PMLR, vol 37, pp 448–456

Jozefowicz R, Zaremba W, Sutskever I (2015) An empirical exploration of recurrent network architectures. In: International conference on machine learning, pp 2342–2350

Luong M-T, Pham H, Manning CD (2015) Effective approaches to attention-based neural machine translation. arXiv:1508.04025

Ordóñez FJ, Roggen D (2016) Deep convolutional and LSTM recurrent neural networks for multimodal wearable activity recognition. Sensors 16(1):115

Osmani A, Hamidi M (2018) Hybrid and convolutional neural networks for locomotion recognition. In: Proceedings of the 2018 ACM UbiComp/ISWC 2018 Adjunct, Singapore, October 08–12, 2018. ACM, pp 1531–1540

Pedregosa F et al (2011) Scikit-learn: machine learning in python. J Mach Learn Res 12:2825–2830

Pham H, Guan M, Zoph B, Le Q, Dean J (2018) Efficient neural architecture search via parameters sharing. In: Dy J, Krause A (eds) Proceedings of the 35th international conference on machine learning, vol 80 of Proceedings of machine learning research, Stockholmsmässan, Stockholm Sweden, 10–15 Jul 2018, PMLR, pp 4095–4104

Plötz T, Hammerla NY, Olivier P (2011) Feature learning for activity recognition in ubiquitous computing. In: International joint conference on artificial intelligence, p 1729

Snoek J, Larochelle H, Adams RP (2012) Practical Bayesian optimization of machine learning algorithms. In: Advances in neural information processing systems, pp 2951–2959

Szegedy C, Liu W, Jia Y, Sermanet P, Reed S, Anguelov D, Erhan D, Vanhoucke V, Rabinovich A (2015) Going deeper with convolutions. In: Proceedings of the IEEE conference on computer vision and pattern recognition, pp 1–9

Szegedy C, Vanhoucke V, Ioffe S, Shlens J, Wojna Z (2016) Rethinking the inception architecture for computer vision. In: Proceedings of the IEEE conference on computer vision and pattern recognition, pp 2818–2826

Wang Z, de Freitas N (2014) Theoretical analysis of Bayesian optimisation with unknown Gaussian process hyper-parameters. arXiv:1406.7758

Yao S et al (2017) Deepsense: a unified deep learning framework for time-series mobile sensing data processing. In: International conference on world wide web, pp 351–360

Zeng M, Nguyen LT, Yu B, Mengshoel OJ, Zhu J, Wu P, Zhang J (2014) Convolutional neural networks for human activity recognition using mobile sensors. In: 2014 6th international conference on mobile computing, applications and services (MobiCASE). IEEE, pp 197–205

Zhong Z, Yan J, Wu W, Shao J, Liu C-L (2018) Practical block-wise neural network architecture generation. In: Proceedings of the IEEE conference on computer vision and pattern recognition, pp 2423–2432

Zoph B, Le QV (2016) Neural architecture search with reinforcement learning. arXiv:1611.01578

Zoph B, Vasudevan V, Shlens J, Le QV (2017) Learning transferable architectures for scalable image recognition. arXiv:1707.07012, 2(6)

Chapter 13
Into the Wild—Avoiding Pitfalls in the Evaluation of Travel Activity Classifiers

Peter Widhalm, Maximilian Leodolter and Norbert Brändle

Abstract Most submissions to the 2018 Sussex-Huawei Locomotion-Transportation (SHL) recognition challenge strongly overestimated the performance of their algorithms in relation to their performance achieved on the challenge evaluation data. Similarly, recent studies on smartphone based trip data collection promise accurate and detailed recognition of various modes of transportation, but it appears that in field tests the available techniques cannot live up to the expectations. In this chapter we experimentally demonstrate potential sources of upward scoring bias in the evaluation of travel activity classifiers. Our results show that (1) performance measures such as accuracy and the average class-wise F1 score are sensitive to class prevalence which can vary strongly across sub-populations, (2) cross-validation with random train/test splits or large number of folds can easily introduce dependencies between training and test data and are therefore not suitable to reveal overfitting, and (3) splitting the data into disjoint subsets for training and test does not always allow to discover model overfitting caused by lack of variation in the data.

13.1 Introduction

Automated trip data collection with smartphones has become a focus of scientific attention. Travel routes and modes of transport can be reconstructed from GPS tracks recorded with a mobile app by analyzing travel speed and proximity to transportation infrastructure (Stenneth et al. 2011; Stopher et al. 2007; Bohte et al. 2008). However, GPS tracking is energy-consuming and can therefore quickly drain the battery of mobile devices. This motivated research towards algorithms to recognize travel

P. Widhalm (✉) · M. Leodolter · N. Brändle
Austrian Institute of Technology, Giefinggasse 2, 1210 Vienna, Austria
e-mail: peter.widhalm@ait.ac.at

M. Leodolter
e-mail: maximilian.leodolter@ait.ac.at

N. Brändle
e-mail: norbert.braendle@ait.ac.at

N. Kawaguchi et al. (eds.), *Human Activity Sensing*,
Springer Series in Adaptive Environments,
https://doi.org/10.1007/978-3-030-13001-5_13

activities with low-power motion sensors such as accelerometer, magnetometer and barometer. Proposed approaches to sensor based Human Activity Recognition report good classification performance for modes of 'active' locomotion such as walking, running and riding a bicycle (Ronao and Cho 2017; Ghosh and Riccardi 2014; Siirtola and Röning 2012). Several studies also report high accuracy for more detailed distinctions between motorized transportation modes such as Car, Bus, Train, and Subway, e.g. Hemminki et al. (2013) with reported 80.1% accuracy for six different transport modes (Manzoni et al. 2010), with 82.14% accuracy for eight transport modes, and Liang and Wang (2017) reporting 94.48% for seven transport modes. However, when available apps and algorithms are tested "in the wild", classification accuracy appears to be much lower (Harding et al. 2017). The Sussex-Huawei Locomotion-Transportation (SHL) recognition challenge (Wang et al. 2018) defined an activity recognition task under simplified conditions, where the training and evaluation data where both collected by a single user and the phone was carried invariably in the front right pocket. The submissions to the SHL challenge included an evaluation of the classification performance conducted by the authors using data that was released along with the correct class labels. The organizers of the challenge ranked the classifiers according to their performance on another data set for which the correct labels were held back. This allowed a direct comparison between the accuracy reported by the authors and the performance achieved in the challenge. Most of the submissions (16 out of 19) report F1 scores above 90%, whereas only two submissions actually achieve this performance on the out-of-sample evaluation data, even though being collected by the same user and phone location. This shows that in most submissions the evaluation conducted by the authors failed to reveal that their models suffered from overfitting.

The experiments in this chapter aim to shed light on undiscovered overfitting and prevalence dependent performance measures as potential sources of bias in the evaluation of transport mode classifiers. The work presented here builds upon our contribution to the SHL recognition challenge presented in Widhalm et al. (2018), where we described our classification algorithm and demonstrated problems related to undiscovered overfitting. Here, we extend our analysis by discussing the role of different performance metrics and using fixed hyper-parameters for classifier training to obtain comparable results across our experiments. We first compare the properties of Accuracy, average F1 score and average Recall regarding their sensitivity to class prevalence in the evaluation data. Using the SHL data set (Gjoreski et al. 2018) we show that the sensor time-series used as input for travel activity recognition are autocorrelated in the sense that temporally close data frames tend to be very similar and that the similarity is a decreasing function of the time lag between them. We also demonstrate the impact of autocorrelation on cross-validation results when using random train-test splits or a large number of folds. Finally, we test the classifier with data of different participants and different carrying positions to show that overfitting due to lack of variation in the data cannot always be discovered by splitting the data into disjoint sub-samples for training and validation.

The remainder of this chapter is organized as follows: Sect. 13.2 describes the data used in our experiments. In Sect. 13.3 we introduce the classification algorithms we used in our experiments. We discuss some commonly used performance metrics in Sect. 13.4. The results of our experiments are reported and discussed in Sect. 13.5, and we wrap up our findings in Sect. 13.6.

13.2 Data

For our experiments we used subsets of the Sussex-Huawei Locomotion (SHL) Dataset presented in Gjoreski et al. (2018). The data was collected with HUAWEI Mate 9 smartphones using the Android application presented in Ciliberto et al. (2017), and contains labels for eight different activity classes: Car, Bus, Train, Subway, Walking, Run, Bike, and Still. The data includes a variety of sensor modalities of which we consider the following sensors for our experiments:

- 3D accelerometer with readings

$$\mathbf{acc}_t = (\mathrm{acc}_{t,x}, \mathrm{acc}_{t,y}, \mathrm{acc}_{t,z}), \tag{13.1}$$

  representing the accelerations along three axes x, y, and z at time t;
- 3D gyroscope with readings

$$\mathbf{gyr}_t = (\mathrm{gyr}_{t,x}, \mathrm{gyr}_{t,y}, \mathrm{gyr}_{t,z}), \tag{13.2}$$

  representing the rate of rotation around the three axes at time t;
- 3D magnetometer with readings

$$\mathbf{mag}_t = (\mathrm{mag}_{t,x}, \mathrm{mag}_{t,y}, \mathrm{mag}_{t,z}), \tag{13.3}$$

  representing the magnetic field strength at time t;
- barometric sensor with readings aap_t representing the ambient air pressure at time t.

All sensors are sampled at 100 Hz. The complete SHL dataset contains data of three users and four different smartphone locations (Hips, Bag, Hand, and Torso).

For our experiments we mostly use the sub-sample released as training data for the SHL recognition challenge. This sub-sample comprises 271.8 h of data collected only by User 1 with the smartphone invariably worn in the front right pocket (corresponding to Hips). However, for testing the classifier with data of different participants and phone locations we also included in our experiments the corresponding other sub-samples. At the time of writing, for the other users and phone locations only a preview data set was available online. This preview includes data from three recording days of all three users and comprises 59 h of annotated recordings for each of the four phone locations.

We also adopt the data pre-processing for the particular classification task of the challenge: the original sensor time series were segmented into frames of 60 s length, and the classification task was to produce for a single 60-s frame a vector of 60×100 class labels, corresponding to the sensor sampling frequency. The classifier had to process each frame without using information from previous or following frames, because in the competition the order of the frames in the test set was randomly permuted.

13.3 Feature Extraction and Classification Algorithms

We preprocess the data to extract features for classifier training as follows: from the 3D accelerometer, gyroscope, and magnetometer time series we compute the magnitudes

$$|\mathbf{acc}_t| = \sqrt{\mathrm{acc}_{t,x}^2 + \mathrm{acc}_{t,y}^2 + \mathrm{acc}_{t,z}^2}, \tag{13.4}$$

$$|\mathbf{gyr}_t| = \sqrt{\mathrm{gyr}_{t,x}^2 + \mathrm{gyr}_{t,y}^2 + \mathrm{gyr}_{t,z}^2}, \tag{13.5}$$

and

$$|\mathbf{mag}_t| = \sqrt{\mathrm{mag}_{t,x}^2 + \mathrm{mag}_{t,y}^2 + \mathrm{mag}_{t,z}^2}. \tag{13.6}$$

Using the 3D accelerometer vector $\mathbf{acc}_t$ we detect events where the device is subject to rotation or jolts (e.g. due to user-interactions with the phone). Rotations are detected by thresholding the variance of the 3D acceleration vector after applying a simple low-pass filter, and jolts are detected by thresholding the variance of the unfiltered acceleration magnitude. As described in Mizell (2003), the gravity component $\mathbf{g}_t$ can be estimated by averaging the acceleration vectors $\mathbf{acc}_t$ over the time interval between pairs of successive rotation events. Estimating the gravity component allows splitting the acceleration vector into the vertical component

$$v_t = \frac{(\mathbf{acc}_t - \mathbf{g}_t)\mathbf{g}_t^\intercal}{|\mathbf{g}_t|} \tag{13.7}$$

and the horizontal component

$$h_t = |\mathbf{acc}_t - v_t \mathbf{g}_t|. \tag{13.8}$$

We extract features from three-second sliding time windows with one second overlap. In particular, we compute

- the minimum, maximum, mean, and standard deviation of $|\mathbf{acc}_t|$, $|\mathbf{gyr}_t|$, $|\mathbf{mag}_t|$, and aap_t,

- the autocorrelation function (ACF) of $|\mathbf{acc}_t|$ and $|\mathbf{mag}_t|$ for time lags from 10 milliseconds to 1 second, and we use the number of ACF zero-crossings and the time lag of the first zero-crossing as features,
- the mean value of v_t and h_t, and
- a binary feature representing whether rotations or jolt events were detected.

We conducted our experiments with the classification algorithm used for our contribution (Widhalm et al. 2018) to the SHL recognition challenge. This algorithm involves two stages. In the first stage the Neural Network computes class posteriors given the above described 'local' data features f_i for each three-second time window. As input for the Neural Network we use vectors $\mathbf{f}_w$ comprising the features f_i, the quadratic terms f_i^2 and the interaction terms $f_i f_j$. In the second stage a Hidden Markov Model uses the results of the first stage as emission probabilities to compute the final classification result. More formally, given the feature vector $\mathbf{f}_w$ of time window w, the neural network estimates the conditional probability distribution $P(c_w|\mathbf{f}_w)$ over the transport mode classes c. As shown in Morgan and Bourlard (1995), an estimate for the scaled emission probabilities

$$\frac{P(\mathbf{f}_w|c_w)}{P(\mathbf{f}_w)} \tag{13.9}$$

can be obtained by dividing the output of the neural network by the relative class frequencies in the training data. The constant scaling factor $P(\mathbf{f}_w)$ can be neglected, as it will not change the classification result. The transition probabilities $P(c_w|c_{w-1})$ can easily be estimated from the training data. The most likely sequence for the 60 s frames of the SHL recognition challenge dataset (c_1, ..., c_{60}) can be computed by maximizing

$$\underset{(c_1,\ldots,c_{60})}{\operatorname{argmax}}\ P(c_1|\mathbf{f}_1) \prod_{w=2..60} P(\mathbf{f}_w|c_w) P(c_w|c_{w-1}) \tag{13.10}$$

using the *Viterbi algorithm*. An illustration of the classification procedure is shown in Fig. 13.1.

Due to the window size (three second sliding window with one second overlap) our classifiers are limited to a resolution of one second, while the desired output consists of 100 class labels per second. Therefore every class label in the sequence is duplicated 100 times to obtain the final result $(c_{1,1}, \ldots, c_{1,100}, \ldots, c_{60,1}, \ldots, c_{60,100})$ where $c_{i,j} = c_i$ for all j.

For the experiments we mostly used the original architecture of the classification algorithm in Widhalm et al. (2018), where the Neural Network had a single hidden layer with 10 units. In this study, however, the model parameters differ slightly, as we train the model without regularization and with fixed convergence criteria and learning rates throughout all experiments to obtain comparable results. We will refer to this classifier as *NN-10*. To demonstrate overfitting, we also use a more complex variant with two hidden layers, the first with 80 units and the second with 60 units. This variant will be referred to as *NN-80-60*.

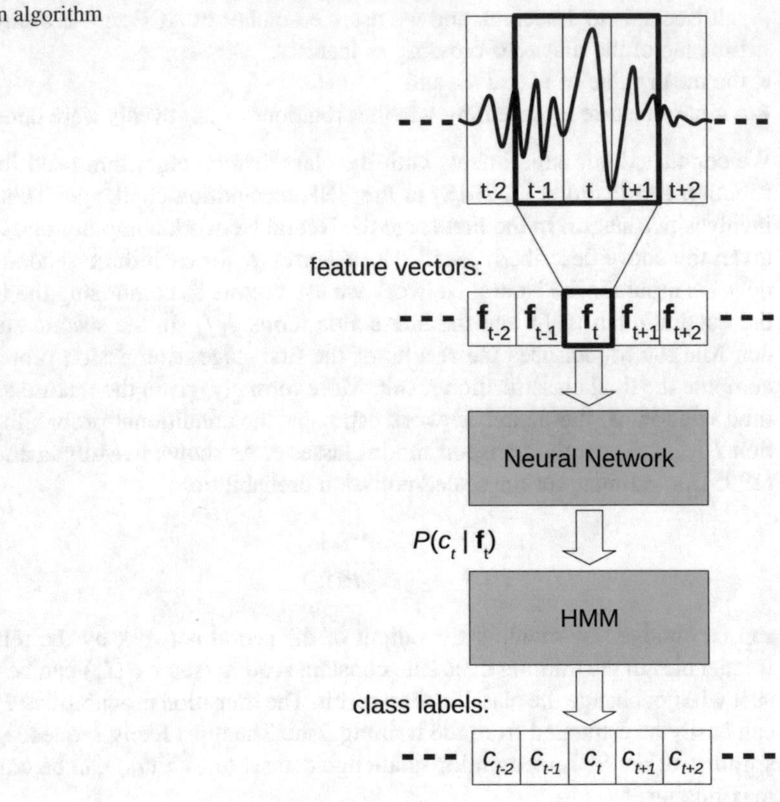

Fig. 13.1 Illustration of the classification algorithm

13.4 Evaluation Metrics

For our experimental analysis we will use some of the most common performance evaluation metrics for classification tasks, namely *Accuracy*, *Recall*, *Precision*, and *F1 score*. Before we describe the individual performance measures in more detail, let us define the confusion matrix **C**, where each entry $c_{i,j}$ is the number of observations of the true class i that were predicted to be class j. Hence, **C** is a square matrix and the numbers in the main diagonal correspond to correct predictions while the off-diagonal entries refer to incorrect predictions.

Accuracy is defined as the fraction of correct classifications across the sample and equally accounts for all classes and predictions:

$$\text{Accuracy} = \frac{\sum_{i=1..N} c_{i,i}}{\sum_{i=1..N} \sum_{j=1..N} c_{i,j}}. \tag{13.11}$$

Hence, Accuracy is an undistorted measure of model performance as long as the class prevalence can be expected to be identical between the population of interest and the evaluation sample. In practice, however, class prevalence can vary strongly across sub-population, and for a particular population of interest it is often unknown. Obviously, prevalence independent performance measures are preferable in such cases. Besides, Accuracy does not allow to distinguish the degree by which the classification model is informed by the data from its general tendency to predict certain classes regardless of the input. For binary classification, a common way to address this problem is using Recall and Precision as complementary performance measures. Binary classification typically aims to distinguish *positive* or *relevant* instances from *negative* or *irrelevant* ones, e.g. medical testing for a disease or retrieval of documents relevant to a query. Recall is defined as proportion of correct positive predictions to the total number of positive instances in the sample, i.e.

$$\text{Recall} = \frac{\text{True Positive}}{\text{True Positive} + \text{False Negative}} = \frac{c_{1,1}}{c_{1,1} + c_{1,2}}, \tag{13.12}$$

and is a prevalence insensitive measure of the model's ability to identify positive instances. In contrast, Precision is the proportion of correct positive predictions to the total number of positive predictions, i.e.

$$\text{Precision} = \frac{\text{True Positive}}{\text{True Positive} + \text{False Positive}} = \frac{c_{1,1}}{c_{1,1} + c_{2,1}}, \tag{13.13}$$

and measures the model's ability to reject negative instances. However, Precision is sensitive to prevalence and is therefore only meaningful when the class prevalence in the population of interest is roughly known, or in particular when positive instances can be considered as rare events in a very large (or even infinite) population. Recall and Precision can be combined by computing their harmonic mean, the so-called F1 score

$$\text{F1} = 2 \times \frac{\text{Precision} \times \text{Recall}}{\text{Precision} + \text{Recall}} \tag{13.14}$$

to obtain a single performance value. Note, however, that neither Recall, Precision or F1 score measures the model's ability to identify negative instances, i.e. the second of the two classes, as it is considered to be *irrelevant*. In a scenario where both classes are equally relevant, a comprehensive performance evaluation therefore has to include analysis of the dual problem with interchanged positive and negative instances. This allows computing the *Inverse Recall* (also called Specificity) and *Inverse Precision*. In a similar fashion, Precision and Recall can be generalized to multi-class problems: the performance measures

$$\text{Recall}_i = \frac{c_{i,i}}{\sum_{j=1..N} c_{i,j}} \tag{13.15}$$

and

$$\text{Precision}_i = \frac{c_{i,i}}{\sum_{j=1..N} c_{j,i}} \tag{13.16}$$

are computed by considering each class at a time as the *positive* or *relevant* label. The class-wise Recall and Precision can be averaged across all classes to obtain

$$\overline{\text{Recall}} = \frac{1}{N} \sum \text{Recall}_i \tag{13.17}$$

and

$$\overline{\text{Precision}} = \frac{1}{N} \sum \text{Precision}_i. \tag{13.18}$$

Similarly, one can also compute class-wise F1 scores, which can be averaged to obtain a single performance value

$$\overline{\text{F1}} = \frac{1}{N} \sum \text{F1}_i \tag{13.19}$$

for all classes. However, we will show in Sect. 13.5.1 that—unlike Recall—the average class-wise F1 score is sensitive to class prevalence and has no clear interpretation.

13.5 Experimental Results

In this section we experimentally demonstrate the following potential sources of bias in performance evaluations with cross-validation: sensitivity to varying class prevalence, autocorrelation, and lack of sufficient variation in the data.

13.5.1 Sensitivity to Class Prevalence

To demonstrate the sensitivity of some popular performance measures to class prevalence, we train model *NN-10* (described in Sect. 13.3) using 75% of the data. From the remaining 25% of the data we draw with replacement to obtain three samples with different class prevalence, one balanced and the other two skewed. The balanced sample consists of 1000 instances of each class. The first of the skewed samples (*skewed sample I*) comprises 5000 instances, respectively, of classes Walk, Car, Subway, and 1000 instances of each of the other classes. The second skewed sample (*skewed sample II*) contains 5000 instances, respectively, of Still, Walk, Run, Bike, and 1000 instances of Car, Bus, Train, and Subway. For each of these samples we compute Accuracy, Recall, Precesion, and F1 score. The results are shown in Table 13.1.

Table 13.1 Evaluation results of classifier *NN-10* using different samples with varying class prevalence

	Still	Walking	Run	Bike	Car	Bus	Train	Subway	Average
Balanced sample:									
Prevalence:	12.5	12.5	12.5	12.5	12.5	12.5	12.5	12.5	
Recall:	94.1	96.4	99.7	99.2	55.4	82.2	82.0	82.5	**86.4**
Precision:	88.1	100.0	99.9	99.6	71.2	64.6	82.2	89.1	**86.8**
F1:	91.0	98.2	99.8	99.4	62.3	72.4	82.1	85.7	**86.3**
Accuracy:									**86.4**
Skewed sample I:									
Prevalence:	5.0	25.0	5.0	5.0	25.0	5.0	5.0	25.0	
Recall:	93.7	96.9	99.6	99.2	55.4	81.4	79.4	83.5	**86.1**
Precision:	72.7	100.0	99.3	99.0	92.1	26.7	53.4	96.8	**80.0**
F1:	81.9	98.4	99.5	99.1	69.2	40.2	63.8	89.7	**80.2**
Accuracy:									**81.6**
Skewed sample II:									
Prevalence:	20.8	20.8	20.8	20.8	4.2	4.2	4.2	4.2	
Recall:	94.1	96.4	99.7	98.9	55.3	80.3	82.4	82.2	**86.2**
Precision:	94.0	100.0	99.8	99.6	64.3	63.7	73.6	83.9	**84.9**
F1:	94.1	98.1	99.7	99.2	59.5	71.0	77.8	83.0	**85.3**
Accuracy:									**93.5**

For the balanced sample the average Recall is identical to Accuracy. The class-wise Recall is similar to the class-wise Precision, i.e. $\text{Recall}_i \approx \text{Precision}_i$, for all classes except Car and Bus, where Precision and Recall differ by roughly the same amount but in opposite directions. As a consequence, the average Recall, average Precision and average F1 score all have similar values. The class-wise Precision values respond strongly to the changed class prevalence in *skewed sample I*. As compared to the balanced sample, Precision is worse for all classes with below-average prevalence and better for classes with increased prevalence. The resulting average Precision, however, is reduced from 86.8% for the balanced sample to only 80.0% in the skewed sample. Accordingly, the average F1 score is reduced from 86 to 80.2%. The changed class prevalence also results in lower Accuracy, since the *skewed sample I* contains a larger fraction of Car and Subway, which are hard for the classifier to distinguish from Bus and Train, respectively.

A large number of instances in *skewed sample II* were drawn from classes the classification model can easily distinguish, which increases Accuracy. On the other hand, average Precision and average F1 score are *lower* than in the balanced sample, because of the lower class-wise Precision of classes with low prevalence. Throughout all three samples Recall remains robust and responds only to the random composition of the samples.

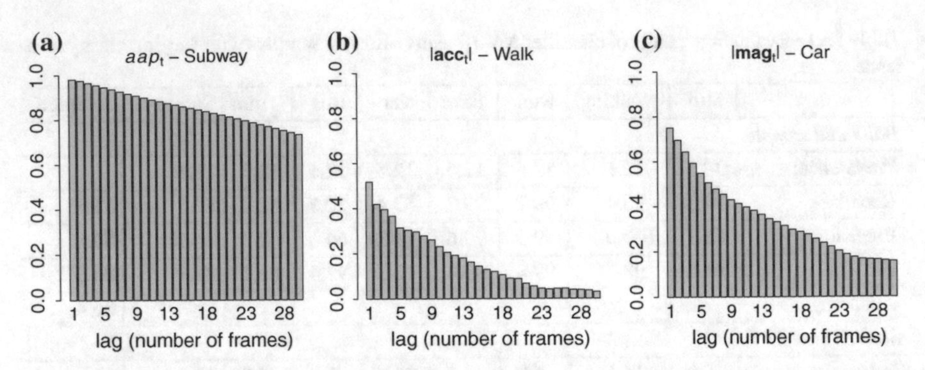

Fig. 13.2 ACF of three different features and activity classes for time lags from 1 to 30 min: **a** the mean of the barometer readings aap_t for class Subway, **b** the standard deviation of the acceleration magnitude $|\mathbf{acc}_t|$ for class Walk, and **c** the mean of the magnetic field strength $|\mathbf{mag}_t|$ for class car

13.5.2 *Cross-Validation with Autocorrelated Data*

We will first show that the input time series data are autocorrelated. To this end we compute the Autocorrelation Function (ACF) of some selected features extracted from the data. To exclude correlations due to the activity class we only consider data of one class at a time. Figure 13.2 shows the ACF of (a) the mean of the barometer readings aap_t for class Subway, (b) the standard deviation of the acceleration magnitude $|\mathbf{acc}_t|$ for class Walk, and (c) the mean of the magnetic field strength $|\mathbf{mag}_t|$ for class Car. The ACF reveals that neighbouring time frames are strongly correlated and should therefore not be split across the samples for training and evaluation. The correlations decrease over time and can presumably be explained by multiple influence factors such as traffic condition, type of road, pavement, walking pace, weather, and altitude.

In our first experiment we demonstrate the potential impact of autocorrelation on the evaluation results. We randomly split the time frames into a training set comprising 75% of the data (12195 frames) and a test set with 25% of the data (4065 frames). This causes temporally neighbouring data frames to be split across the samples for training and evaluation. The resulting average F1-score is $\overline{\text{F1}}$ =91% for classifier *NN-10* and $\overline{\text{F1}}$ =95.8% for classifier *NN-80-60*. The evaluation details are given in Tables 13.2 and 13.3. For comparison we now order the samples temporally and *backtest* the classifier, using the first 75% as 'past' data for model training and the last 25% as 'future' data to evaluate the model. This minimizes the undesirable correlations between the two samples as only few correlated time frames are split among training and test data. We perform 4-fold cross-validation by repeating the experiment four times, each time using a different quarter of the data for testing. Here,

Table 13.2 Cross-validation with random train-test splits: classifier *NN-10*

		Predicted activity								
	(%)	Still	Walking	Run	Bike	Car	Bus	Train	Subway	Average
Actual activity	Still	13.4	0.0	0.0	0.1	0.0	0.1	0.3	0.1	
	Walking	0.2	12.8	0.0	0.1	0.0	0.0	0.0	0.1	
	Run	0.0	0.0	4.5	0.0	0.0	0.0	0.0	0.0	
	Bike	0.1	0.0	0.0	12.2	0.0	0.1	0.0	0.0	
	Car	0.4	0.0	0.0	0.0	13.4	0.6	0.2	0.0	
	Bus	0.3	0.0	0.0	0.0	1.4	11.4	0.2	0.0	
	Train	0.9	0.0	0.0	0.0	0.3	0.1	13.8	1.0	
	Subway	0.2	0.0	0.0	0.0	0.0	0.0	2.6	8.7	
Recall:		95.6	96.5	99.3	97.8	91.1	85.8	85.5	75.0	**90.8**
Precision:		85.9	98.9	99.1	97.2	88.7	93.0	80.4	88.2	**91.4**
F1:		90.5	97.7	99.2	97.5	89.9	89.2	82.9	81.1	**91.0**

Table 13.3 Cross-validation with random train-test splits: classifier *NN-80-60*

		Predicted activity								
	(%)	Still	Walking	Run	Bike	Car	Bus	Train	Subway	Average
Actual activity	Still	13.6	0.1	0.0	0.1	0.0	0.0	0.2	0.2	
	Walking	0.1	13.2	0.0	0.0	0.0	0.0	0.0	0.0	
	Run	0.0	0.0	4.2	0.0	0.0	0.0	0.0	0.0	
	Bike	0.1	0.0	0.0	12.4	0.0	0.0	0.0	0.0	
	Car	0.1	0.0	0.0	0.0	14.7	0.5	0.1	0.0	
	Bus	0.1	0.0	0.0	0.0	0.4	12.4	0.1	0.0	
	Train	0.4	0.0	0.0	0.0	0.1	0.1	13.9	0.7	
	Subway	0.2	0.0	0.0	0.0	0.0	0.0	1.9	9.9	
Recall:		96.4	98.0	99.5	98.6	95.5	94.4	91.3	82.5	**94.5**
Precision:		93.1	98.7	99.3	98.9	96.5	95.2	85.9	90.8	**94.8**
F1:		94.7	98.4	99.4	98.8	96.0	94.8	88.5	86.5	**94.6**

classifier *NN-10* achieves $\overline{\text{F1}}$ =86.5% and the performance of the more complex model *NN-80-60* is only $\overline{\text{F1}}$ =82.4%. Note that model *NN-80-60* performs worse than model *NN-10*, although with random train test splits it seemed to be superior. This result shows that cross-validation with random train-test splits is biased and not suitable for detecting model overfitting. Moreover, it should be noted that even in the improved cross-validation procedure described above the proportion of correlated instances in the test set increases multiplicatively with the number of folds, as the size of the test set decreases but the number of problematic instances is constant and determined by the ACF of the data. Therefore, a large number of folds (or even a leave-one-out scheme) can also lead to biased evaluation results.

13.5.3 Cross-Validation with Data Lacking Sufficient Variation

With the following experiment we show that overfitting due to lack of variation in the data cannot be discovered with cross-validation, i.e. by using disjoint sub-samples of the same data set for training and validation, respectively. The data set used for the SHL recognition challenge contained only one user who carried the smartphone invariably in the front right pocket. This reduces variability and can cause the classification model to overfit. Cross-validation measures classification performance *given* the limited variability in the available data. In contrast, we now test the classification performance of model *NN-10* with the SHL preview data set of different users and device locations. While the average F1-score with cross-validation and *backtesting* was $\overline{\mathbf{F1}}$ =86.5%, the classifier achieves only $\overline{\mathbf{F1}}$ =78.6% when tested with data of the same user but with different device locations (Hand, Torso, and Bag). Using data from the same device location (Hips) but different users (User 2 and User 3) the evaluation result is $\overline{\mathbf{F1}}$ =79.0%, and when both device location and user is different the classifier achieves only $\overline{\mathbf{F1}}$ =72.6%. The results of this experiment are detailed in Tables 13.4 13.5 and 13.6.

Table 13.4 Backtesting results of classification model *NN-10*

		Predicted activity								
	(%)	Still	Walking	Run	Bike	Car	Bus	Train	Subway	Average
Actual activity	Still	13.6	0.1	0.0	0.1	0.0	0.0	0.2	0.2	
	Walking	0.1	13.2	0.0	0.0	0.0	0.0	0.0	0.0	
	Run	0.0	0.0	4.2	0.0	0.0	0.0	0.0	0.0	
	Bike	0.1	0.0	0.0	12.4	0.0	0.0	0.0	0.0	
	Car	0.1	0.0	0.0	0.0	14.7	0.5	0.1	0.0	
	Bus	0.1	0.0	0.0	0.0	0.4	12.4	0.1	0.0	
	Train	0.4	0.0	0.0	0.0	0.1	0.1	13.9	0.7	
	Subway	0.2	0.0	0.0	0.0	0.0	0.0	1.9	9.9	
Recall:		96.4	98.0	99.5	98.6	95.5	94.4	91.3	82.5	**86.5**
Precision:		83.3	98.6	98.1	96.6	79.6	79.5	73.2	87.9	**87.1**
F1:		88.9	97.0	98.6	97.1	77.2	77.5	78.2	77.8	**86.5**

Table 13.5 Backtesting results of classification model *NN-80-60*

		Predicted activity								
	(%)	Still	Walking	Run	Bike	Car	Bus	Train	Subway	average
Actual activity	Still	11.9	0.1	0.0	0.1	0.1	0.1	1.5	0.2	
	Walking	0.3	12.7	0.0	0.3	0.0	0.0	0.1	0.1	
	Run	0.0	0.0	4.1	0.1	0.0	0.0	0.0	0.0	
	Bike	0.1	0.1	0.2	12.3	0.0	0.0	0.0	0.0	
	Car	0.4	0.0	0.0	0.0	10.7	2.8	1.1	0.2	
	Bus	0.1	0.0	0.0	0.0	1.8	10.4	0.4	0.1	
	Train	0.8	0.0	0.0	0.0	0.4	0.1	12.9	1.3	
	Subway	0.2	0.0	0.0	0.0	0.1	0.1	2.8	8.9	
Recall:		85.3	94.6	97.9	96.5	70.4	80.9	83.4	73.6	**85.3**
Precision:		86.1	97.6	95.3	96.1	82.2	77.2	68.7	82.9	**85.8**
F1:		85.7	96.1	96.6	96.3	75.8	79.0	75.3	78.0	**85.3**

Table 13.6 Results of classification model *NN-10* with data of varying persons and phone locations

(%)	Still	Walking	Run	Bike	Car	Bus	Train	Subway	Average
User 1, Hand/5Torso/Bag:									
Recall:	85.4	66.5	93.2	94.8	49.3	80.8	83.0	72.7	**78.2**
Precision:	83.0	98.5	99.4	68.3	56.6	77.0	78.4	83.6	**80.6**
F1:	84.2	79.4	96.2	79.4	52.7	78.8	80.6	77.7	**78.6**
User 2+3, Hips:									
Recall:	93.0	63.8	86.3	97.0	37.6	82.3	86.7	78.5	**78.1**
Precision:	87.0	95.0	99.6	72.7	56.9	80.5	73.1	91.5	**82.0**
F1:	89.9	76.3	92.4	83.1	45.3	81.4	79.3	84.5	**79.0**
User 2+3, Hand/Torso/Bag:									
Recall:	90.7	53.5	80.1	66.7	48.5	77.6	84.5	78.5	**72.5**
Precision:	85.6	96.6	94.6	54.5	57.1	60.1	73.1	79.3	**75.1**
F1:	88.1	68.9	86.7	60.0	52.4	67.7	78.4	78.9	**72.6**

13.6 Conclusion

With our experiments we identified undiscovered overfitting and prevalence sensitive evaluation metrics as potential reasons for biased evaluation.

Although Accuracy and the F1 score are popular performance measures, we showed that both metrics are sensitive to class prevalence, which complicates their interpretation and hinders comparison of the results obtained with different samples. In particular, optimizing a classification model for a test sample with specific class prevalence does not guarantee optimal performance for a different data sample. On the other hand, Recall is insensitive to class prevalence. Moreover, the average Recall is identical to Accuracy when class prevalence is balanced and therefore has a clear and intuitive interpretation.

In order to obtain reliable cross-validation results it is crucial that the data is split into independent samples for model training and evaluation. For the case of travel activity classification with sensor time series we demonstrated that this assumption can easily be violated by random train-test splits or large number of cross-validation folds, because the data is autocorrelated. We showed how to mitigate this problem with backtesting, which minimizes the distribution of correlated instances across the samples used for training and evaluation.

Our experiments also demonstrated that if the available data lacks variation, cross-validation cannot reliably assess the model's ability to generalize to unseen future data. In practice, however, it is very challenging to collect a data sample that covers a wide range of variation, because the recorded smartphone sensor signals result from a multitude of factors besides the travel model: the individual style of driving and locomotion (e.g. driving profile, gait), the travel route (e.g. road class and pavement), the traffic condition, the particular type of vehicle, interactions with the smartphone (gaming, texting, calls), and the phone carrying position. Our future research will therefore aim to define efficient data collection and cross-validation strategies to achieve with minimal effort the required variation for reliable estimates of the future performance in the wild.

References

Bohte W, Maat K, Quak W (2008) A method for deriving trip destinations and modes for GPS-based travel surveys. Res. Urbanism Ser. 1(1):127–143

Ciliberto M, Morales FJO, Gjoreski H, Roggen D, Mekki S, Valentin S (2017) High reliability android application for multidevice multimodal mobile data acquisition and annotation. In: Proceedings of the 15th ACM conference on embedded network sensor systems, p. 62. ACM

Ghosh A, Riccardi G (2014) Recognizing human activities from smartphone sensor signals. In: Proceedings of the 22nd ACM international conference on Multimedia, pp 865–868. ACM (2014)

Gjoreski H, Ciliberto M, Wang L, Morales FJO, Mekki S, Valentin S, Roggen D (2018) The university of sussex-huawei locomotion and transportation dataset for multimodal analytics with mobile devices. IEEE Access. https://doi.org/10.1109/ACCESS.2018.2858933

Harding C, Srikukenthiran S, Zhang Z, Habib KN, Miller E (2017) On the user experience and performance of smartphone apps as personalized travel survey instruments: results from an experiment in Toronto. In: The paper presented at the 11th international conference on transport survey methods, Quebec

Hemminki S, Nurmi P, Tarkoma S (2013) Accelerometer-based transportation mode detection on smartphones. In: Proceedings of the 11th ACM Conference on Embedded Networked Sensor Systems, p 13. ACM (2013)

Liang X, Wang G (2017) A convolutional neural network for transportation mode detection based on smartphone platform. In: 2017 IEEE 14th international conference on mobile Ad Hoc and sensor systems (MASS), pp 338–342. IEEE

Manzoni V, Maniloff D, Kloeckl K, Ratti C (2010) Transportation mode identification and real-time CO_2 emission estimation using smartphones. Massachusetts Institute of Technology, SENSEable City Lab

Mizell D (2003)Using gravity to estimate accelerometer orientation. In: Null, p 252. IEEE

Morgan N, Bourlard HA (1995) Neural networks for statistical recognition of continuous speech. Proceedings of the IEEE 83(5):742–772

Ronao CA, Cho SB (2017) Recognizing human activities from smartphone sensors using hierarchical continuous hidden Markov models. Int J Distribut Sensor Netw 13(1):1550147716683687

Siirtola P, Röning J (2012) Recognizing human activities user-independently on smartphones based on accelerometer data. IJIMAI 1(5):38–45

Stenneth L, Wolfson O, Yu PS, Xu B (2011) Transportation mode detection using mobile phones and gis information. In: Proceedings of the 19th ACM SIGSPATIAL international conference on advances in geographic information systems, pp 54–63. ACM

Stopher P, FitzGerald C, Xu M (2007) Assessing the accuracy of the Sydney household travel survey with GPS. Transportation 34(6):723–741

Wang L, Gjoreskia H, Murao K, Okita T, Roggen D (2018) Summary of the sussex-huawei locomotion-transportation recognition challenge. In: Proceedings of the 2018 ACM international joint conference and 2018 international symposium on pervasive and ubiquitous computing and wearable computers, pp 1521–1530. ACM

Widhalm P, Leodolter M, Brändle N (2018) Top in the lab, flop in the field?: evaluation of a sensor-based travel activity classifier with the SHL dataset. In: Proceedings of the 2018 ACM international joint conference and 2018 international symposium on pervasive and ubiquitous computing and wearable computers, pp 1479–1487. ACM (2018)

Chapter 14
Effects of Activity Recognition Window Size and Time Stabilization in the SHL Recognition Challenge

Michael Sloma, Makan Arastuie and Kevin S. Xu

Abstract The Sussex-Huawei Locomotion-Transportation (SHL) recognition challenge considers the problem of human activity recognition from inertial sensor data collected at 100 Hz from an Android smartphone. We propose a data analysis pipeline that contains three stages: a pre-processing stage, a classification stage, and a time stabilization stage. We find that performing classification on "raw" data features (i.e. without feature extraction) over extremely short time windows (e.g. 0.1 s of data) and then stabilizing the activity predictions over longer time windows (e.g. 15 s) results in much higher accuracy than directly performing classification on the longer windows when evaluated on a 10% hold-out sample of the training data. However, this finding *does not hold* on the competition test data, where we find that accuracy drops with decreasing window size. Our submitted model uses a random forest classifier and attains a mean F1 score over all activities of about 0.97 on the hold-out sample, but only about 0.54 on the competition test data, indicating that our model does not generalize well despite the use of a hold-out sample to prevent test set leakage.

14.1 Introduction

Sensor-based human activity recognition is a long-studied problem in ubiquitous computing (Bao and Intille 2004; Ravi et al. 2005; Choudhury et al. 2008; Lara and Labrador 2013; Hammerla et al. 2016; Gjoreski et al. 2018) and is a key challenge to creating context-aware computing. The objective of the Sussex-Huawei Locomotion-Transportation (SHL) recognition challenge was to recognize 8 differ-

M. Sloma · M. Arastuie · K. S. Xu (✉)
University of Toledo, Toledo, OH 43606, USA
e-mail: Kevin.Xu@utoledo.edu

M. Sloma
e-mail: Michael.Sloma@rockets.utoledo.edu

M. Arastuie
e-mail: Makan.Arastuie@rockets.utoledo.edu

N. Kawaguchi et al. (eds.), *Human Activity Sensing*,
Springer Series in Adaptive Environments,
https://doi.org/10.1007/978-3-030-13001-5_14

Table 14.1 Our 3-stage data analysis pipeline. Classification stage uses either of the 2 classifiers

Pre-processing	Classification	Time stabilization
1. Replace missing values 2. Apply `MinMaxScaler` 3. Reshape to smaller window size	• Random forest • Deep convolutional neural network	1. Repeat predicted labels to obtain prediction for each sample (at 100 Hz) 2. Stabilize predicted classes over *l* seconds

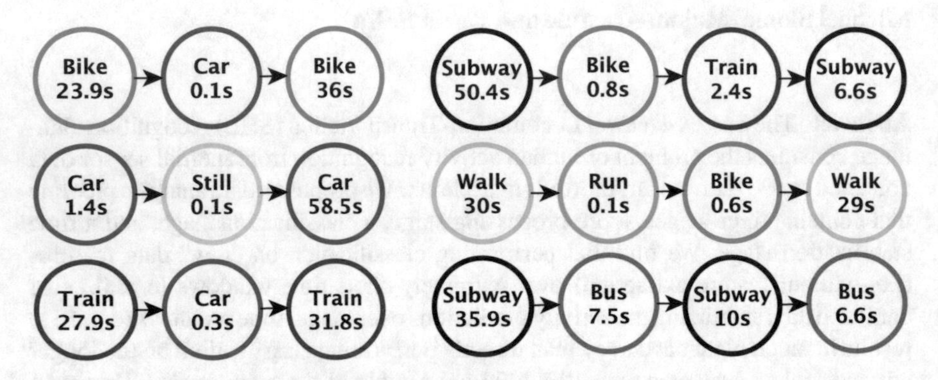

Fig. 14.1 Examples of inconsistencies in predicted activities from the competition test data using a classifier without time stabilization

ent types of activities from a variety of sensor data collected at 100 Hz using a smartphone (Gjoreski et al. 2018; Wang et al. 2018). In the SHL recognition challenge, the 8 activities corresponded to 8 different modes of locomotion and transportation: still, walk, run, bike, car, bus, train, and subway.

Our approach to the SHL recognition challenge consists of a three-stage data analysis pipeline: a pre-processing stage, a classification stage, and a time stabilization stage, as shown in Table 14.1. In the pre-processing stage, we split up the sensor data into short time windows and normalize each sensor's readings so that they are in the same range. The normalized readings for each sensor over the short time windows are concatenated to form a "raw" feature vector that is input to a classification algorithm (e.g. random forest) along with the dominant activity in the time window, which is treated as the class label. The classification algorithm predicts a class (activity) for the time window, treating each individual window independently.

We input the predicted activities from each time window into a *time stabilization* stage to correct predicted classes that are inconsistent with plausible behavior, such as a prediction of train for 27.9 s, followed immediately by 0.3 s of car, and then going back to train for the rest of the 1-min window. More examples of these inconsistencies for predicted activities are shown in Fig. 14.1. We consider two different time stabilization approaches: a majority vote-based approach that allows changes in activity predictions only once every *l* seconds and a more flexible time stabilization heuristic that allows changes at arbitrary times provided that a predicted activity persists for at least *l* seconds.

We experiment with two types of classifiers, random forests and deep convolutional neural networks, and different window sizes for both the pre-processing and time stabilization stages. During the challenge, labels for the competition test data were not available, so we evaluated our performance by using a 10% hold-out sample from the training data. This chapter extends our previous work (Sloma et al. 2018) with analysis of our results on the competition test data, now that labels have been made available.

Our submitted model uses a random forest classifier on 0.1-s windows with predictions stabilized over 15-s windows. On the hold-out sample, we find that performing classification on such short time windows and then stabilizing activity predictions over longer time windows results in much higher accuracy than directly performing classification on raw data features over the longer windows. However, this finding *does not hold* on the competition test data, where we find that accuracy drops with decreasing window size. The model attains a mean F1 score over all activities of about 0.97 on the hold-out sample, but only about 0.54 on the competition test data, indicating that our model does not generalize well despite the usage of a hold-out sample to prevent test set leakage. Our results illustrate one of the most difficult challenges in activity recognition—generalizing high recognition accuracy in the lab to actual performance in the field.

14.2 Data Description

The SHL dataset (Gjoreski et al. 2018) includes a total of 366 h of recorded data, 271 h for training and 95 h for testing. During the challenge, only activity labels were available for the training data. Activity labels for the testing data were made available after the challenge. This data was collected from a single person using Huawei Mate 9 smartphone, worn in the front right pants pocket. Data from the following sensors are available: accelerometer (3-axis), gyroscope (3-axis), magnetometer (3-axis), linear acceleration (3-axis), gravity (3-axis), orientation (quaternions), and ambient pressure. Some of the sensor streams, such as linear acceleration and gravity, are "software sensors" derived from other hardware sensor streams. Each sensor recorded data at 100 Hz. The data was segmented into 1-min windows, with 6,000 samples each. Labels of activities were provided for each of the 6,000 samples in the window.

The frequency of each activity in the training data is shown in Fig. 14.2a, where each occurrence corresponds to a single 0.01-s sample. The activities appear to be pretty uniformly distributed within the training set, with the exception of running. The number of transitions between activities are shown in Fig. 14.2b, where the direction of edge indicates a transition from the activity at the head to the activity at the tail. Most transitions appear to and from the activities Still and Walk, as one might expect, with the most frequent transition being from Walk to Still (75 occurrences). The frequencies of activities and transitions in the test data are shown in Fig. 14.2c and d, respectively. The distribution of activities is more imbalanced than the training set, and there are also much more frequent transitions between activities Still, Bus, and Walk.

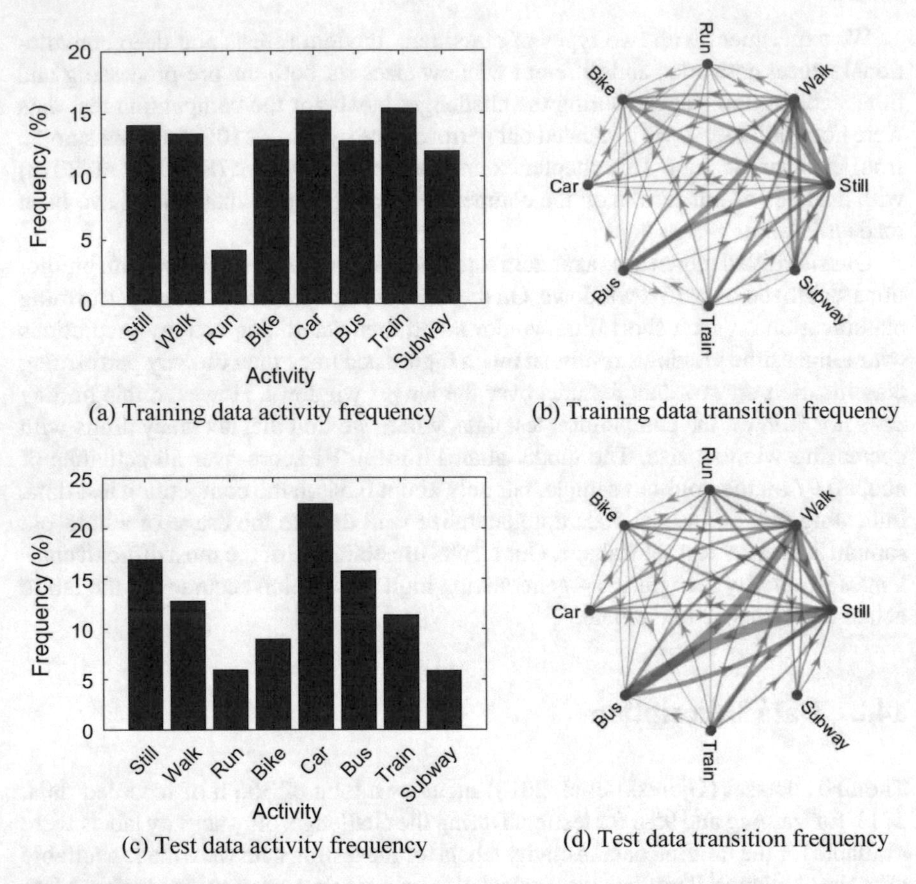

Fig. 14.2 Frequencies of activities and mid-window transitions between activities in the competition training data (**a**), (**b**) and testing data (**c**), (**d**). Thicker edges in the directed graphs denote more frequent transitions

14.3 Methods

Our data analysis pipeline consists of three stages: a pre-processing stage, a classification stage, and a time stabilization stage. Unlike most prior work on classification-based methods for activity recognition (Bao and Intille 2004; Ravi et al. 2005; Choudhury et al. 2008; Lara and Labrador 2013; Gjoreski et al. 2018), we do not use a feature extraction stage, so we feed the "raw" data features into the classifiers.

14.3.1 Data Pre-processing

In several cases, we observed missing data in a sensor stream. In these cases, we replace the missing data by the mean value of the sensor reading in the respective

1-min window. We then stack the data from the 20 sensors into a tensor of dimensions $(N, 6000, 20)$, where N denotes the number of training examples. We then normalize each feature (sensor) to an approximate range of $[-1, 1]$ using the `MinMaxScaler` from scikit-learn (Pedregosa et al. 2011) using the `partial_fit` function.

Changes in activity could occur at any time within the 1-min windows, resulting in a class change mid-window. This inspired us to use window sizes smaller than the provided size of 1 min to enable classification of activities at finer resolutions. Some initial analysis showed that, in the 16,310 training windows, there are 651 places where a class change occurs, so creating windows shorter than 1 min could improve classification accuracy.

We reshape the data to fit the desired window size. For example, using 10-s windows, each window would be represented by a matrix of 1000 samples $\times$ 20 sensors. We then restack the windows to form a tensor of size $(6N, 1000, 20)$. We reshape the labels in a similar fashion and then perform majority vote to obtain a single activity label for each window. Creating shorter time windows allows for finer time resolution in activity labels compared to using 1-min windows because a single activity label is applied to the entire window size.

14.3.2 Classification Algorithms

Since we do not have a feature extraction stage in our pipeline, we place a significant demand on the classification algorithm to identify activities from the "raw" sensor data features. Thus, we consider two classification algorithms capable of learning complex class structures: random forests and deep convolutional neural networks.

14.3.2.1 Random Forest

A random forest is a classifier consisting of a collection of tree-structured classifiers, where each tree casts a vote for the most popular predicted class for an input example (Breiman 2001). Each tree is independently constructed using a bootstrap sample of the data set. At every internal node of each tree, a random subset of features is considered for the split at that node.

Random forests are easy to tune for two reasons. First, there are only two main parameters: the number of trees in the forest and the number of features used at every split. Second, the model is not very sensitive to these parameters, and once a set of reasonable parameters are found, which will most likely happen after a few trials, the performance is going to be close to the model with the best possible parameters.

We use the scikit-learn v0.19.1 (Pedregosa et al. 2011) implementation of random forests which adds a few extra parameters to the model, namely the maximum depth of the trees as well as the split criterion. Moreover, in contrast to Breiman's original publication (Breiman 2001), the scikit-learn implementation combines classifiers by averaging their predicted probabilities over classes (activities), instead of letting each classifier vote for a single class.[1]

14.3.2.2 Deep Convolutional Neural Networks

Neural networks have been used for years for modeling complex non-linear relationships. Deep neural networks (DNNs) contain many more layers than traditional neural networks, and unlike traditional machine learning algorithms, have the ability to automatically learn features from raw data (LeCun et al. 2015). This is especially advantageous because we choose to feed the raw sensor data (with minimal preprocessing) into a classification algorithm rather than first extracting features. Deep learning has been successfully used for human activity recognition using wearables (Hammerla et al. 2016) and thus shows promise for classifying activities in the SHL recognition challenge.

For this challenge, we make use of 1-D (temporal) convolutional neural network (CNN) layers. Deep CNNs (Krizhevsky et al. 2012) comprise of convolutional layers interleaved with max-pooling layers, outputting to a series of traditional (dense) neural network layers, and finally outputting to a softmax layer. The aim of a CNN approach is to use locality and time "closeness" to help gain context from the surrounding data to identify the activity that is occurring at a given time. This also allows for some amount of translational invariance for the patterns detected in each window.

Our deep CNN framework uses Keras (Chollet 2015) with a TensorFlow backend. All the deep CNNs we tested contain at least 3 temporal convolutional layers, at least one max-pooling layer, and at least one fully connected layer. The final output layer uses a softmax activation function across the 8 possible classes (activities). The convolutional layers vary in kernel width, number of filters, and the stride width. The max-pooling layers find the maximum value within their pool width, which adds translational invariance to the model. The max-pooling layers vary in their pool width through the layers in the model. Rectified linear unit (ReLU) activations are used for the output of each convolutional layer, max-pooling layer and fully connected layer, except for the final fully connected layer, which uses a softmax. We apply dropout regularization to all the fully connected layers. The architecture of the best performing model is shown in Table 14.2.

14.3.3 Time Stabilization

Slicing every 1-min window to 600 0.1-s windows will make the algorithm more prone to detect every sudden and sporadic movement and falsely labeling it as its own separate activity. For instance, this can be stopping for a couple of seconds while biking, or standing up in a lengthy train ride to stretch. Although these indeed are

[1] http://scikit-learn.org/stable/modules/ensemble.html#forest.

Table 14.2 Best performing deep CNN (uses 1-s time windows)

Layer name	Layer description			
	Kernel size	Filters	Stride	Activation
Conv1	50	256	1	relu
Conv2	25	256	1	relu
Max1	–	–	2	–
Conv3	10	256	1	relu
Conv4	2	256	1	relu
Flatten				
	Nodes	Dropout rate		Activation
Dense1	2048	0.1		relu
Dense2	2048	0.1		relu
Dense3	2048	0.1		relu
Dense4	2048	0.1		relu
Dense5	2048	0.1		relu
Dense6	2048	0.1		relu
Dense7	2048	0.1		relu
Dense8	2048	0.1		relu
Dense9	1024	0.1		relu
Dense10	8	0		softmax

short changes in activities, we do not count them as switching to another activity, which is what we are aiming to detect. Thus, we decided to regulate these changes in a post-processing step to stabilize the classifier outputs (predictions) of neighboring time windows.

14.3.3.1 Majority Vote

Our first approach is a majority vote. We set the prediction for all samples of an l-second window to the most predicted class within that window. Note that the majority vote window length l must be at least the length of the window size used for classification, as its purpose is to smooth out the predicted classes over time. One obvious shortcoming of this method is that, if $l = 10$ s and the activity changes (due to an actual change or just misclassification) between two 10-s windows, then there is a high chance of mislabeling one or both of the two consecutive windows.

14.3.3.2 Time Stabilization Heuristic

This shortcoming of the majority vote approach leads to the design of a more flexible time stabilization heuristic which accounts for mid-window changes in the activity prediction. In this approach, similar to the majority vote, we need to decide how long an activity should last before considering it an actual switch from one activity to another. That is going to be the l parameter in this approach, which we denote as the *stable window length*, and it must also be at least the length of the classification window size, similar to majority vote.

For every 1-min window, scanning from the beginning, we find the first class that is predicted for l consecutive seconds. Denote this class by S. Next, starting from the first predicted class, if predictions do not persist for l seconds, then we change them to S; however, if a specific predicted class lasts for at least l seconds, then we set S to that class to represent a mid-window change in activity. If the 1-min window ends with a prediction that lasts for less than l seconds, then we leave this prediction unchanged because we simply do not know whether or not the predicted class is going to persist after the window ends because the order of time windows in the competition test set was not made available in the challenge. Finally, if there is no stable prediction within the entire window, then we resort back to a majority vote on the entire window.

14.4 Experiments

To test our data analysis pipeline during the challenge, we split the competition training data using an 80/20 stratified split that preserves the distributions of the classes (activities) in the two sets. The split was based on the majority vote class label of the whole 1-min window. We use the 80% to train our classifiers. We further split the 20% into equally sized validation and hold-out sets. The validation set is used for training the deep CNN and for tuning hyperparameters for both classifiers. We then apply our final choices of models to the hold-out data, which we use as a "simulated" test set during the challenge itself, with labels for the competition test data not yet available. This split into validation and hold-out data allows us to avoid test set leakage, which is often a source of overfitting.

Now that the challenge has concluded, we present two sets of results. We first show results for our classifiers evaluated on the hold-out data, which comprises 10% of the competition training data. This is the evaluation we used during the challenge. We then show results for our classifiers evaluated on the competition test data now that labels are available. The two evaluation settings are shown in Fig. 14.3.

In order to evaluate accuracy, we repeat every predicted class for the entire window to produce a prediction for each sample (0.01 s). This process would result in an $N \times 6000$ matrix (same size as the original label matrix) which we use to calculate the F1 score for each class. We use the individual F1 scores for the classes as well as the mean F1 score over all classes, which is the actual metric used for evaluation in the challenge, as our accuracy metrics while tuning our models.

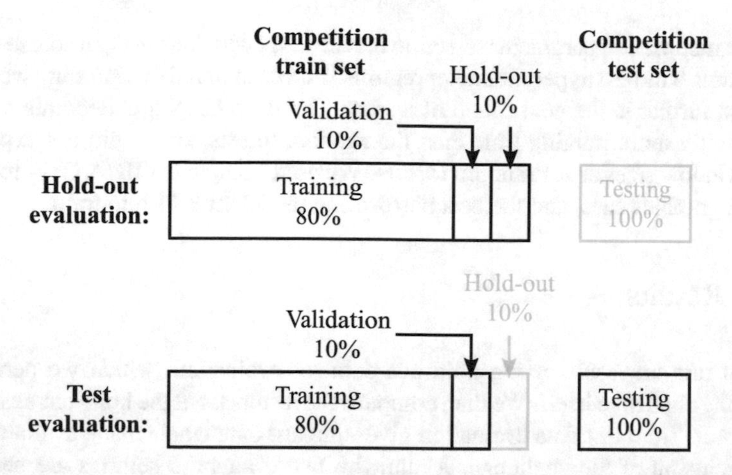

Fig. 14.3 Illustration of evaluation settings for our data analysis pipeline. The hold-out evaluation uses only the competition train set, while the test evaluation uses both the competition train and test sets

14.4.1 Random Forest

As mentioned in Sect. 14.3, random forests are not very sensitive to their hyperparameters so we did not spend too much time on parameter tuning. We use Gini impurity as the splitting criterion. Next, for the number of trees in the forest (`n_estimators`), and their maximum depth (`max_depth`), we start out with two small numbers, 50 and 10, respectively. As we increased these two parameters, we started seeing improvements in the performance until the improvement stopped as we went above 500 trees with 150 as their maximum depth. The other parameter to tune was the number of features to be selected at random at each split (`max_features`). We experimented with using the square root and log of the number of possible features as well as considering all features at every split and consistently got the best performance when we utilized the square root number of possible features. The scikit-learn package also offers a few more parameters which were all set to their default values. This led to our final model parameters: `n_estimators = 500`, `max_depth = 150`, `max_features = 'sqrt'`, and `criterion = 'gini'`.

This particular model was trained and tested on a Linux server with 128 GB of RAM and 2 Intel Xeon processors (E5-2697 v3 @ 2.60 GHz) with a total of 28 CPU cores. However, only 22 cores and ~50 GB of RAM were utilized during the training process, which took about 3 h. Predicting the test data took ~48 s plus ~2 s for the time stabilization.

14.4.2 Deep Convolutional Neural Networks

Since our model had two more parameters than in a typical setting for deep neural networks (window size and stable window length l), we focused more on tuning those two parameters. The parameter selection for the CNN was done using a coarse-to-fine grid search. The best hyperparameter regions from each round of searching were then explored further in the next round of testing. Our deep CNN architectures required significantly more training time than the random forests, so we did not explore as many window sizes as for random forests. We used a single NVIDIA GeForce GTX 1080 Ti graphics card, and the best performing model took 11 h to train.

14.5 Results

We first present results using hold-out data for evaluation, which we performed during the challenge itself. We then compare the results from the hold-out evaluation with results from test data evaluation given the test data labels made available after the conclusion of the challenge. Again, the two evaluation settings are shown in Fig. 14.3.

14.5.1 Hold-out Data Evaluation

14.5.1.1 Effects of Window Size

We first present results from our classifiers without any time stabilization, shown in Table 14.3. For both our random forest and deep CNN models, the average F1 score increases as we shrink the window size, with F1 score finally decreasing when we shrink the windows to 0.05 s. This is most likely due to the fact that, as the window size decreased, there were more samples to train on; however, each sample was slowly losing context due to the shorter window.

Another interesting observation here is that the F1 scores for various activities seem to reach their peak at different window sizes. For instance, walking and running are best predicted at a 1-s window size, while most other activities seem to do better when the window size is shorter. One interpretation of this behavior might be that for walking and running the sensors' values have more variation over time, thus looking at a longer window gives us a better chance at correctly classifying them.

14.5.1.2 Effects of Time Stabilization

As shown in Table 14.4a, the simple majority vote increased the F1-score from 0.952 to 0.971 when the majority vote is taken from the entire 1-min window. In addition, the accuracy seems to increase with increasing majority vote window length, reaching a maximum for majority voting using the entire 1-min window, which would not allow any changes of predicted activities within each minute. However, as discussed in Sect. 14.2, we know that there are indeed changes in activities within the 1-min frames. Thus, we employed our time stabilization heuristic to allow mid-window activity changes.

Table 14.3 Effects of window size on F1 scores of our classifiers without time stabilization on hold-out data

(a) Random forest

Activity	Window size (s)					
	60	10	1	0.25	0.1	0.05
Still	0.893	0.923	0.946	0.955	0.957	**0.958**
Walk	0.887	0.951	**0.964**	0.963	0.961	0.960
Run	0.822	0.978	**0.986**	**0.986**	0.983	0.978
Bike	0.917	0.938	0.961	0.964	**0.964**	0.963
Car	0.910	0.935	0.957	0.968	0.972	**0.973**
Bus	0.800	0.869	0.899	0.922	**0.928**	0.927
Train	0.808	0.875	0.918	0.932	0.934	**0.936**
Subway	0.808	0.872	0.897	0.910	**0.914**	0.913
Average	0.856	0.918	0.941	0.950	**0.952**	0.951

(b) Deep CNN

Activity	Window size (s)	
	10	1
Still	0.917	**0.963**
Walk	0.971	**0.975**
Run	0.987	**0.991**
Bike	0.955	**0.970**
Car	0.967	**0.980**
Bus	0.895	**0.937**
Train	0.908	**0.955**
Subway	0.887	**0.948**
Average	0.936	**0.965**

With our time stabilization heuristic, as seen in Table 14.4b, we were able to achieve a higher F1 score of 0.974 at a smaller stable window length (10 s). Another interesting observation is that, just like initial window size selection, different activities seem to be best predicted with different stable window sizes. Therefore, it is possible for a more restricted heuristic that uses different stable window sizes for different activities to achieve a higher average F1 score. Moreover, it is worth noting that when the stable window length is set to 60 s, the results are identical to the majority vote.

Since time stabilization is a heuristic, we decided that any predicted activity that lasts for less than 15 s should not be considered its own activity. Thus, we conservatively set the stable window size to 15 s, despite the fact that 10 s achieved a slightly higher average F1 score. Besides, a 10-s stable window allows for very frequent activity changes, which may not be particularly realistic. The time stabilization step is very fast and requires no training on the actual data, so changing the stable window size is easy and does not cause any overhead in computation.

Table 14.4 Effects of time stabilization window length *l* on the F1 scores of random forest classifier predictions on 0.1-s windows with (a) majority vote and (b) time stabilization heuristic on hold-out data

(a) Majority vote

Activity	Window length (s)					
	60	30	15	10	5	1
Still	0.962	**0.965**	0.964	0.963	0.962	0.959
Walk	0.970	**0.978**	0.977	0.977	0.977	0.969
Run	0.984	**0.992**	0.991	**0.992**	0.991	0.989
Bike	0.985	**0.989**	0.981	0.982	0.980	0.971
Car	**0.988**	0.985	0.980	0.980	0.977	0.973
Bus	**0.961**	0.948	0.943	0.941	0.937	0.931
Train	**0.956**	0.954	0.948	0.948	0.940	0.937
Subway	**0.962**	0.943	0.942	0.937	0.928	0.918
Average	**0.971**	0.969	0.966	0.965	0.962	0.956

(b) Time stabilization heuristic

Activity	Window length (s)					
	60	30	15	10	5	1
Still	0.962	0.964	0.966	**0.968**	0.966	0.962
Walk	0.970	0.973	0.979	**0.981**	0.977	0.976
Run	0.984	0.984	0.984	0.985	0.985	**0.989**
Bike	0.985	0.985	**0.989**	**0.989**	0.984	0.976
Car	**0.988**	**0.988**	0.986	0.986	0.981	0.975
Bus	0.961	**0.963**	0.962	0.960	0.944	0.935
Train	0.956	0.959	0.958	**0.960**	0.946	0.940
Subway	0.962	**0.964**	0.961	0.963	0.943	0.922
Average	0.971	0.973	0.973	**0.974**	0.966	0.959

We observe similar results when applying our time stabilization heuristic to our deep CNNs, as shown in Table 14.5. Applying time stabilization increased the F1 score for every single activity. Due to the significant computation time involved in training our deep CNNs, we did not arrive at this result until after the challenge deadline, so our submitted data analysis pipeline during the challenge used random forest on 0.1-s windows with time stabilization using a 15-s stable window size.

14.5.2 Test Data Evaluation

The test data provided by the challenge consists of over 34 million samples to be predicted. Our predictions using our random forest model with time stabilization yielded the activity frequencies shown in Fig. 14.4. If we reconstruct the predictions into the provided 1-min windows, there are a total of 5,698 windows, out of which we predicted that 4,981 do not have any transitions between activities.

Table 14.5 Effects of time stabilization (with the stable window length l set to 15 s) on F1 scores of our deep CNN models for 1-s time windows on hold-out data

Activity	Time stabilization	
	Without	With
Still	0.963	**0.979**
Walk	0.975	**0.989**
Run	0.991	**0.993**
Bike	0.970	**0.990**
Car	0.980	**0.992**
Bus	0.937	**0.967**
Train	0.955	**0.969**
Subway	0.948	**0.979**
Average	0.965	**0.982**

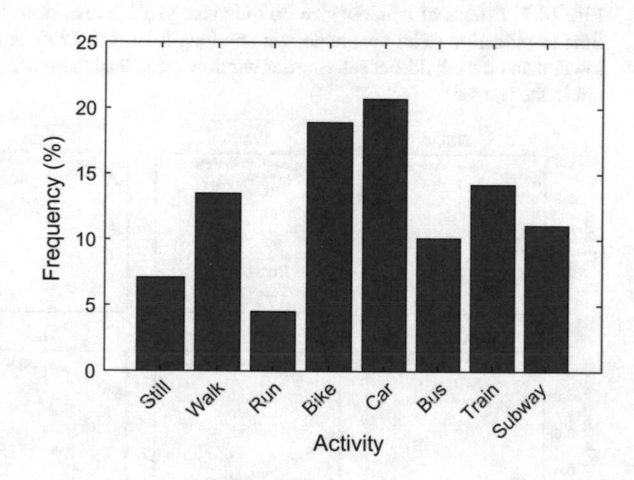

Fig. 14.4 Predicted activity frequency on competition test data. Still and Bus are under-represented while Bike and Subway are over-represented compared to actual activity labels shown in Fig. 14.2c

In our prediction, the total number of transitions between two activities within a 1-min window was 717. Most of these transitions were due to the fact that the time stabilizing heuristic does not stabilize the predictions at the end of the 1-min windows, since these windows were not in order in the provided test set. If we ignore that fact and stabilize end of each window as well, we will end up only with 141 transitions, which better resembles the training data. Most of these 717 transitions are at least feasible in reality. A few examples would be running for 30 s, still for 26 s, then start running again, or the same scenario but with biking instead of running.

The effects of window size on the accuracy of our random forest classifiers is shown in Fig. 14.5. First of all, notice that the average test set F1 score for 0.1-s windows is 0.538, which is significantly lower than the 0.969 for the hold-out set.[2]

[2]F1 scores for the hold-out set may not match results in Sect. 14.5.1 exactly due to a different 80/20 random stratified split being used.

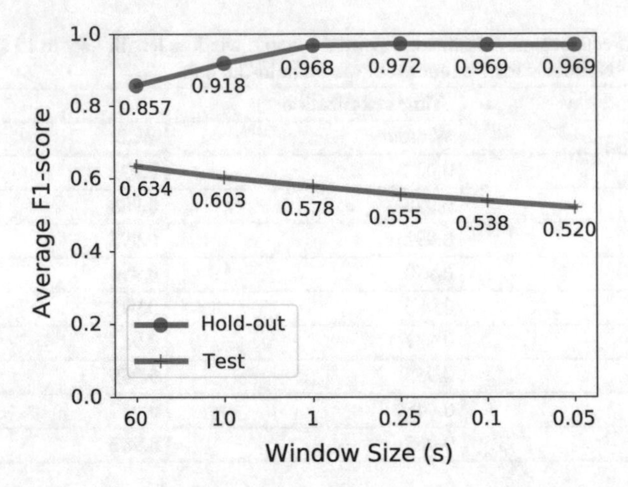

Fig. 14.5 Effects of window size on the average F1 scores of our random forest classifiers with time stabilization (with the stable window length l set to 15 s). Accuracy on the test set is much lower than on the hold-out set. Shorter window sizes lead to higher accuracy in the hold-out set but not in the test set

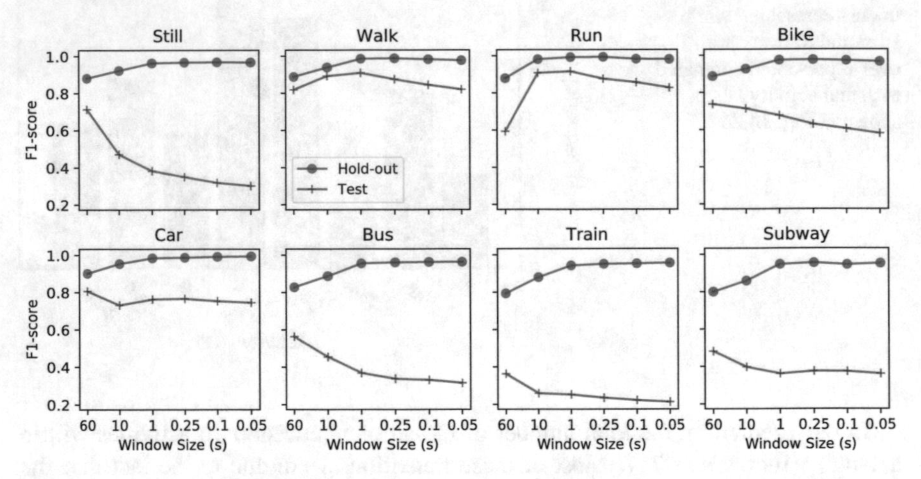

Fig. 14.6 Effects of window size on the F1 scores of individual activities for our random forest classifiers with time stabilization (with the stable window length l set to 15 s). Still, Bus, Train, and Subway are poorly classified in the test set, particularly for short window sizes

Thus, it appears that our model did not generalize well, despite the use of the hold-out set to prevent test set leakage. Notice also that the average F1 score on the test set appears to decrease monotonically with decreasing window sizes, unlike with the hold-out set. Thus, a window size of 60 s results in the worst hold-out set accuracy but the best test set accuracy. When we examine the test set F1 scores for individual activities, as shown in Fig. 14.6, we notice a significant degradation in accuracy of predicting the Still class as we shorten the window size. Bus, Train, and Subway were also extremely poorly predicted compared to the hold-out set while walk and run did not suffer as much.

In all cases, we found that time stabilization improved the accuracy of our random forest classifiers, as shown in Fig. 14.7. Our proposed time stabilization heuristic leads to a greater average F1 score improvement than majority vote for all stable window lengths l in both the hold-out and test evaluation settings (except for $l = 60$ s for which the two stabilization techniques are equivalent). Thus, despite the degradation in classifier accuracy, we find that the performance improvement attributable to our time stabilization heuristic is maintained and even slightly improved in the test evaluation setting.

Our findings for the random forest classifiers apply also to our deep CNN models, as shown in Table 14.6. The decrease in accuracy from hold-out to test evaluation is even more dramatic than for the random forest classifiers, with a drop in average F1 score of about 0.5. We see poor prediction of the Still, Bus, Train, and Subway activities in the test set, similar to the random forest classifiers. Our time stabilization heuristic again improves the accuracy of the classifier predictions, in this case by 0.023, which is consistent with the improvement on the predictions from the random forest classifiers.

14.6 Discussion

Overall, we were able to attain very accurate activity predictions in both of our models on the hold-out set, with F1 scores above 0.96 after time stabilization. However, this high level accuracy did not translate to the competition test set for either model, with F1 scores around 0.5. Among all 8 activities, Bus, Train, and Subway proved to be the 3 hardest activities to predict on the hold-out set. All 3 are high speed activities with irregular stops. A person can be still in a train or subway or can be walking around, and so on. Thus, given their nature it is also not surprising that our time stabilization heuristic provoked the highest improvement on these three activities resulting in an F1 score of around 0.96. On the competition test set, our model also did an extremely poor job of predicting when the person is still, whereas walking and running suffered only small drops in F1 score.

We found that some sensors proved to be more predictive than the others. Shrinking the window down to 0.1 s resulted in a random forest model with 200 features, that is 10 samples from each of the 20 sensors. In order to observe how important a sensor was in our model, we computed the average Gini importance of each feature over all trees in the ensemble, and then for each sensor we summed up all of its samples' feature importance measures, which resulted in the values shown in Fig. 14.8. Although the pressure sensor has the highest importance, this is due to the other sensors having multiple axes. If we sum up the importance measures across all axes for each sensor, then the top 3 most important sensors in our model, in the order of importance, end up being the gravity sensor, magnetometer, and pressure.

Fig. 14.7 Effects of stable window length l of time stabilization heuristic (TS) compared to majority vote (MV) on the F1 score of our random forest classifiers (with 0.1 s time window) in both hold-out and test evaluation settings. Our proposed time stabilization heuristic leads to a greater average F1 score improvement than majority vote

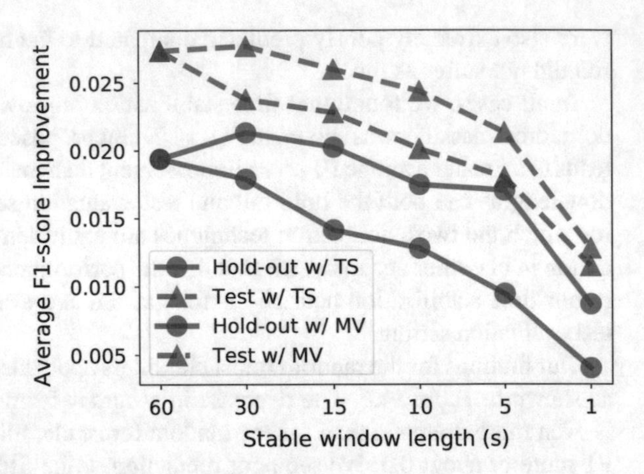

Table 14.6 Effects of (a) window size and (b) time stabilization on F1 scores of our deep CNN models on test data. Our time stabilization heuristic was applied to the 1-s time windows with stable window length l set to 15 s

(a) Window sizes

Activity	Window size (s)	
	10	1
Still	**0.114**	0.109
Walk	**0.823**	0.794
Run	0.733	**0.893**
Bike	0.377	**0.516**
Car	**0.589**	0.570
Bus	0.226	**0.282**
Train	0.142	**0.178**
Subway	0.292	**0.338**
Average	0.412	**0.460**

(b) Time stabilization

Activity	Time stabilization	
	Without	With
Still	**0.114**	0.103
Walk	0.823	**0.841**
Run	0.733	**0.939**
Bike	0.377	**0.553**
Car	**0.589**	0.585
Bus	0.226	**0.305**
Train	0.142	**0.187**
Subway	0.292	**0.353**
Average	0.460	**0.483**

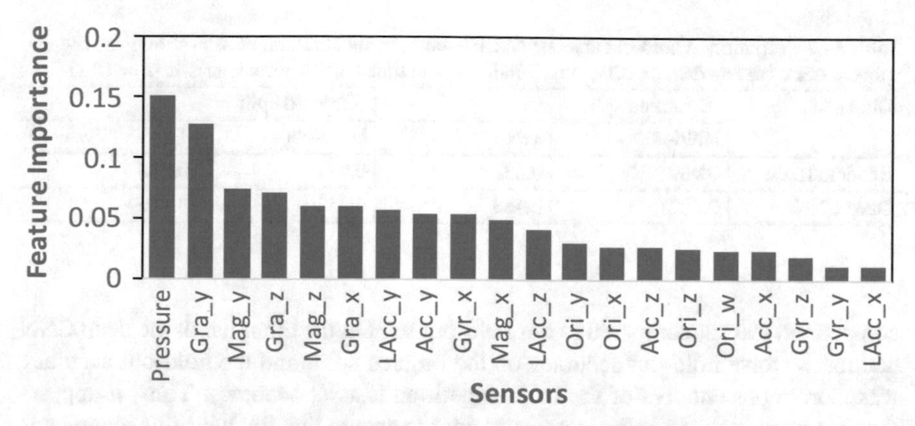

Fig. 14.8 The sensor importance for the random forest model with 0.1 s time windows. Feature importance values for each sensor are added to obtain the total importance of a sensor

Furthermore, the rationale behind using a random forest model on small intervals followed by the time stabilization heuristic instead of simply taking larger intervals, comes down to one main reason: increasing the number of training examples. It is well-known that increasing the number of training examples will generally increase the accuracy of a classifier. On the other hand, small windows have two disadvantages. First, we start losing surrounding context, and second, we may pick up on many sudden movements and label them as their own activity. In order to address the first issue, we tried out a series of different window sizes and finally settled on 0.1 s since we were still seeing improvements as we were shrinking the window size until we went below 0.1 s. Next, to mediate the false classification of sudden movements, we designed the time stabilization heuristic. This way we have the advantage of having access to a larger number of training examples while minimizing the effects of its drawbacks. Unfortunately, the improvements that we observed in the hold-out set did not carry over to the test set, for which longer windows provided more accurate predictions.

In this work, we opted to split data into train and test sets using a random 80/20 stratified split to preserve the distributions of activities in the two sets. Prior work has suggested that an ordered split, i.e. first ordering time windows and then splitting into train and test sets based on contiguous chunks of time, provides a more realistic evaluation of activity recognition performance in the field (Hammerla and Plötz 2015; Widhalm et al. 2018; Wang et al. 2018). The downside of such an approach is that it no longer guarantees that the distributions of different activities in the train and test sets is balanced.

We re-ran our proposed data pipeline (including re-training classifiers) using an 80/20 ordered rather than random stratified split to see if we observe the upward bias mentioned in Hammerla and Plötz (2015), Widhalm et al. (2018). As shown in Table 14.7, the results are mixed. The random forest classifier obtained roughly the same result for both types of splits, with a substantial drop in the F1 score on the

Table 14.7 Variation in hold-out and test data F1 scores using 80/20 random stratified and ordered splits of train and test data on 0.1-s time windows with time stabilization heuristic ($l = 15$ s)

Classifier	Stratified split		Ordered split	
	Hold-out	Test	Hold-out	Test
Random forest	0.969	0.537	0.978	0.538
Deep CNN	0.982	0.483	0.453	0.466

competition test set compared to our hold-out set. On the other hand, the deep CNN had much worse hold-out accuracy on the ordered split, and the hold-out accuracy was more representative of the the competition test set accuracy. Thus, it appears that just using ordered splits is not sufficient to ensure that the hold-out accuracy is representative of actual recognition accuracy in the field. This is an issue worthy of further investigation, particularly due to its high practical importance.

Acknowledgements We would like to thank the University of Toledo's Office of Undergraduate Research for providing funding for Michael Sloma through the Undergraduate Summer Research and Creative Activities Program (USRCAP).

References

Bao L, Intille SS (2004) Activity recognition from user-annotated acceleration data. In: Proceedings of the 2nd international conference on pervasive computing, pp 1–17

Breiman L (2001) Random forests. Mach Learn 45(1):5–32

Chollet F et al (2015) Keras. https://keras.io

Choudhury T, Borriello G, Consolvo S, Haehnel D, Harrison B, Hemingway B, Hightower J, Klasnja P, Koscher K, LaMarca A, Landay JA, LeGrand L, Lester J, Rahimi A, Rea A, Wyatt D (2008) The mobile sensing platform: an embedded activity recognition system. IEEE Pervasive Comput 7(2):32–41

Gjoreski H, Ciliberto M, Wang L, Morales FJO, Mekki S, Valentin S, Roggen D (2018) The University of Sussex-Huawei locomotion and transportation dataset for multimodal analytics with mobile devices. IEEE Access 6:42592–42604

Hammerla NY, Halloran S, Plötz T (2016) Deep, convolutional, and recurrent models for human activity recognition using wearables. In: Proceedings of the 25th international joint conference on artificial intelligence, pp 1533–1540

Hammerla NY, Plötz T (2015) Let's (not) stick together: pairwise similarity biases cross-validation in activity recognition. In: Proceedings of the ACM international joint conference on pervasive and ubiquitous computing, pp 1041–1051

Krizhevsky A, Sutskever I, Hinton GE (2012) ImageNet classification with deep convolutional neural networks. Adv Neural Inf Process Syst 25:1097–1105

Lara OD, Labrador MA (2013) A survey on human activity recognition using wearable sensors. IEEE Commun Surv Tutor 15:1192–1209

LeCun Y, Bengio Y, Hinton G (2015) Deep learning. Nature 521(7553):436

Pedregosa F, Varoquaux G, Gramfort A, Michel V, Thirion B, Grisel O, Blondel M, Prettenhofer P, Weiss R, Dubourg V, Vanderplas J, Passos A, Cournapeau D, Brucher M, Perrot M, Duchesnay E (2011) Scikit-learn: machine learning in Python. J Mach Learn Res 12:2825–2830

Ravi N, Dandekar N, Mysore P, Littman ML (2005) Activity recognition from accelerometer data. In: Proceedings of the 17th conference on innovative applications of artificial intelligence, pp 1541–1546

Sloma M, Arastuie M, Xu KS (2018) Activity recognition by classification with time stabilization for the SHL recognition challenge. In: Proceedings of the ACM international joint conference on pervasive and ubiquitous computing adjunct, pp 1616–1625

Wang L, Gjoreskia H, Murao K, Okita T, Roggen D (2018) Summary of the Sussex-Huawei locomotion-transportation recognition challenge. In: Proceedings of the ACM international joint conference on pervasive and ubiquitous computing adjunct. ACM, pp 1521–1530

Widhalm P, Leodolter M, Brändle N (2018) Top in the lab, flop in the field?: evaluation of a sensor-based travel activity classifier with the SHL dataset. In: Proceedings of the ACM international joint conference on pervasive and ubiquitous computing adjunct. ACM, pp 1479–1487

Chapter 15
Winning the Sussex-Huawei Locomotion-Transportation Recognition Challenge

Vito Janko, Martin Gjoreski, Gašper Slapničar, Miha Mlakar, Nina Reščič, Jani Bizjak, Vid Drobnič, Matej Marinko, Nejc Mlakar, Matjaž Gams and Mitja Luštrek

Abstract The Sussex-Huawei Locomotion-Transportation Recognition Challenge presented a unique opportunity to the activity-recognition community to test their approaches on a large, real-life benchmark dataset with activities different from those typically being recognized. The goal of the challenge was to recognize eight locomotion activities (Still, Walk, Run, Bike, Car, Bus, Train, Subway). This chapter describes the submissions winning the first and second place. They both start with data preprocessing, including a normalization of the phone orientation. Then, a wide set of hand-crafted domain features in both frequency and time domain are computed and their quality evaluated. The second-place submission feeds the best features into an XGBoost machine-learning model with optimized hyper-parameters, achieving the accuracy of 90.2%. The first-place submission builds an ensemble of models, including deep learning models, and finally refines the ensemble's predictions by smoothing with a Hidden Markov model. Its accuracy on an internal test set was 96.0%.

The first two authors should be regarded as joint first authors.

V. Janko (✉) · M. Gjoreski (✉) · G. Slapničar · M. Mlakar · N. Reščič · J. Bizjak · V. Drobnič · M. Marinko · N. Mlakar · M. Gams · M. Luštrek
Department of Intelligent Systems, Jožef Stefan Institute, Ljubljana, Slovenia
e-mail: vito.janko@ijs.si

V. Janko · M. Gjoreski · N. Reščič · J. Bizjak · M. Gams · M. Luštrek
Jožef Stefan Postgraduate School, Ljubljana, Slovenia

N. Kawaguchi et al. (eds.), *Human Activity Sensing*,
Springer Series in Adaptive Environments,
https://doi.org/10.1007/978-3-030-13001-5_15

15.1 Introduction

Smart devices have become an indispensable part of our lives. Smartphones, smart watches and other wearables accompany us everywhere. The grand vision of ubiquitous computing is that these devices will know as much as possible about our context in order to provide the best possible service, and contribute to our safety, health, comfort and overall quality of life.

What we are doing at any given moment is a key element of our context, which is why activity recognition (AR) is intensely researched. Most research on AR is focused on our bodies, dealing with activities such as walking, sitting and lying. However, since we spend a lot of time in vehicles—transportation studies show that the average commute time is up to 80 min a day (Olsson et al. 2013) and we also travel for other purposes—this is probably an area that deserves more attention. If the grand vision of ubiquitous computing is to be realized, our devices should know not only whether we are walking or sitting, but also whether riding a train or driving a car.

The Sussex-Huawei Locomotion-Transportation (SHL) dataset consists of seven months of recordings of smartphone sensors during eight modes of locomotion. It was collected to develop methods for AR, traffic analysis, localization, sensor fusion and other problems. These methods can support mobile services such as travel and traffic advice, adapting phone operation to the mode of locomotion (notifications, volume, Wi-Fi and GPS…), using this mode in games, for music selection and a myriad of other purposes application developers will think of. Most importantly for this chapter, the dataset provides an excellent challenge for ubiquitous computing researchers—to apply existing and new methods for locomotion AR on more data than most of them are able to collect on their own and of course to take on the SHL Challenge (Wang et al. 2018).

This chapter describes two approaches to locomotion AR that placed first and second at the SHL Challenge. The first one is classical, based on over a decade of experience in the AR field. It starts with preprocessing the orientation data, continues with extracting a large number of expertly crafted features and selecting the best of them, and finishes with feeding the features into a classification model with tuned hyper-parameters. The second approach builds multiple models, some of which are trained with deep learning algorithms. The outputs of the models are combined into an ensemble, and the final predictions are smoothed with a Hidden Markov model (HMM). In designing both approaches, we applied the principle of multiple knowledge (Gams 2001): used different "viewpoints" (feature categories, models) and sensibly combined them (via feature selection, in an ensemble), with the expectation that the result will be superior to using single (high-quality) viewpoints. The two approaches were originally described in two HASCA workshop papers (Janko 2018; Gjoreski 2018). They are described in this chapter as a single approach with several improvements, particularly regarding smoothing, and post-competition commentary is added.

15.2 Related Work

The AR domain has been thoroughly explored in the past using body-worn sensors, ambient sensors and combinations of the two. Here we focus only on body-worn sensors, since smartphones and smart watches are most frequently used for AR today.

The most frequent AR task is classifying activities related to movement, e.g., walking, running, standing still and cycling (Kozina et al. 2013). Different approaches using standard machine learning (ML) and feature extractions have been used and tested on various datasets (Hung 2014; Ronao and Cho 2016; Roggen 2010; Teng 2018; Janko et al. 2017; Cvetković 2017). However, deep learning attempts are becoming increasingly prevalent (Gjoreski 2016). Surprisingly, this domain is still not dominated by deep learning, unlike computer vision and some other domains, most likely because deep learning in AR is not clearly superior to classical ML.

Several attempts have been made in the past towards classification of just one activity, or distinguishing between activities related to one domain (e.g. transportation) (Ravì et al. 2017; Wang et al. 2010; Reddy et al. 2010). However, the SHL Challenge seems to be more ambitious, trying to classify a wide variety of activities both human movement- and transportation-related. Therefore, the main and most important related work to this chapter are other approaches submitted to this challenge, some of which are included in this book.

15.3 Dataset

The SHL Dataset used in this work is publicly available and thoroughly described (Gjoreski et al. 2018). The subset used for the SHL Challenge was recorded with a Huawei Mate 9 smartphone carried inside the front right pocket (not fixed orientation) by a single participant over a period of four months for 5–8 h per day.

The data comes from a variety of sensors in the smartphone: accelerometer, gyroscope, magnetometer, linear accelerometer, gravity, orientation (expressed with quaternions) and barometer; all sampled with the frequency of 100 Hz. The data is labeled with eight classes: Still, Walk, Run, Bike, Car, Bus and Subway. A class label was applied to each sensor sample. In aggregate, around 266 h of labeled data was provided. The goal of the competition was to train a model on this data, and then use it to classify an additional 100 h of unlabeled test data.

The provided data was split into 1-min intervals, which were then shuffled—with the original order for the labeled data provided as part of the dataset description. Since consecutive intervals can display substantial similarities (e.g., the phone may be in exactly the same orientation), it is essential that the dataset is placed in the original order before it is split into train, validation and test set. If this is not done, one of consecutive intervals is often placed in the train and the other in the validation/test set, making some irrelevant features (e.g., specific phone orientation) appear valuable for the prediction and thus resulting in overfitting.

We used 50% of the labeled data for training, 25% for validation and 25% for testing. After the parameters of the most accurate ML model were fixed, a different distribution—90% for training and 10% for testing—was used to get an insight on how the behavior of the model changes if given more training data. For training of the final model, all the data was used. In all cases, we verified that the activity distribution remained similar in all listed splits.

15.4 Features

In order to apply most ML methods constituting our approach, the data had to be preprocessed and some features describing each signal extracted.

15.4.1 Preprocessing

First, the data was downsampled from 100 to 50 Hz. This significantly reduced the computational load of the subsequent calculations, while not significantly affecting the classification accuracy. The additional practical benefit of the downsampling lies in reduced consumption of the phone battery, should the system be used in a real-time setting.

Second, "virtual" sensor streams were calculated based on the real ones. These sensor streams had the same frequency and were used in the same manner as the original data streams for calculating features. They can be grouped in three categories:

- *Magnitudes*. Sensors with three axes had the magnitude ($m = \sqrt{x^2 + y^2 + z^2}$) calculated, which was used in addition to the sensor streams of individual axes.
- *De-rotated sensors.* These were computed by de-rotating acceleration and magnetometer data from body (phone) coordinate system to the North-East-Down (NED) coordinate system by multiplying them with the rotation matrix R_{NB} created with quaternions $\left[q_w, q_x, q_y, q_z\right]$ as given below. This was done in order to obtain orientation-independent sensor information, which is important to avoid overfitting to specific orientations of the phone. However, since orientation information can also be relevant, both groups of sensor streams were retained.

$$R_{NB} = \begin{bmatrix} 1 - 2(q_y^2 + q_z^2) & 2(q_x q_y - q_w q_z) & 2(q_x q_z + q_w q_y) \\ 2(q_x q_y + q_w q_z) & 1 - 2(q_x^2 + q_z^2) & 2(q_y q_z - q_w q_x) \\ 2(q_x q_z - q_w q_y) & 2(q_y q_z + q_w q_x) & 1 - 2(q_x^2 + q_y^2) \end{bmatrix}$$

$$\begin{bmatrix} x \\ y \\ z \end{bmatrix}_N = R_{NB} \begin{bmatrix} x \\ y \\ z \end{bmatrix}_B$$

- *Roll, pitch and yaw.* Finally, quaternions were used to compute Euler angles—roll, pitch and yaw. Orientation is usually presented with quaternions to avoid the *gimbal lock* point of singularity, however, Euler angles are better for extracting features since each of them has a clear real-world meaning, whereas quaternion components are only really meaningful when taken all together. Because of that, we only used Euler angles in the subsequent steps.

$$\begin{aligned} roll &= \arcsin\left(2\left(q_w q_y + q_z q_x\right)\right) \\ pitch &= \arctan\left(\frac{2\left(q_w q_x + q_y q_z\right)}{1 - 2\left(q_x q_x + q_y q_y\right)}\right) \\ yaw &= \arctan\left(\frac{2\left(q_w q_z + q_x q_y\right)}{1 - 2\left(q_y q_y + q_z q_z\right)}\right) \end{aligned}$$

In the end, counting all separate sensor axes, we worked with 30 different data streams.

15.4.2 Feature Extraction

Features were extracted from 1-min windows of data. This window size was chosen as it was the largest possible given the limitations imposed by the nature of the competition, and we achieved the highest classification accuracy using it. Since transitions between activities are rare, long windows did not have a significant negative impact on the performance. Labels were calculated for each window as the most frequent label in that window.

Nearly all features were extracted from each individual data stream. Three categories of features were computed. First, domain features that have proven themselves in our previous work in similar domains (Cvetković et al. 2015, 2018) including in a previously won competition (Kozina et al. 2013). These features are described in the work by Cvetković et al. (2018).

Second, we tried to generate all features using the tsfresh library (Tsfresh 2018). While the library is capable of generating a large number of features, is seemed too slow given the size of our dataset. Consequently, we only generated some features that seemed interesting and were not included in other categories—namely minimum, maximum, autocorrelation, number of samples above/below the mean and the average difference between two sequential data samples.

Finally, we calculated some features from the frequency domain, which describe the periodicity of the signals. These features were calculated using power spectral density (PSD), which is based on the fast Fourier transform (FFT). PSD characterizes the frequency content of a given signal and can be estimated using several techniques. The simplest one is to use a periodogram, which is obtained by taking the squared-magnitude of the FFT components. An alternative to periodogram is the Welch's method, which is also widely used and commonly considered superior to periodogram. It differs from a traditional periodogram in the fact that it computes the average of the periodograms of multiple overlapping segments of the signal to reduce the variance of the PSD. In our work, we opted to use the Welch's method to obtain the PSD.

Using the PSD is only suitable when the signal is clearly periodic. As we chose to use a 1-min window, any periodic pattern in the signals is successfully captured, as shown in Fig. 15.1.

We implemented the frequency-domain features as given in related work (Su et al. 2014). Some were slightly modified or expanded in accordance with our expert knowledge. The following features were computed.

- *Three largest magnitudes.* Three peaks with the largest magnitude from the PSD were considered. These tell us the dominant frequencies in the signal. Both the magnitude values and the frequencies (in Hz) were taken as features.
- *Energy.* Calculated as the sum of the squared FFT component magnitudes. The energy was then normalized by dividing it with the window length.

$$energy = \frac{1}{N}\sum_{n=0}^{N-1}|x(n)|^2$$

  where $x(n)$ is the n-th FFT component and N is the window length.
- *Entropy.* Calculated as the information entropy of the normalized FFT component magnitudes. It helps discriminating between activities with similar energy features.

$$entropy = -\sum_{n=0}^{N-1} x(n)\log(x(n))$$

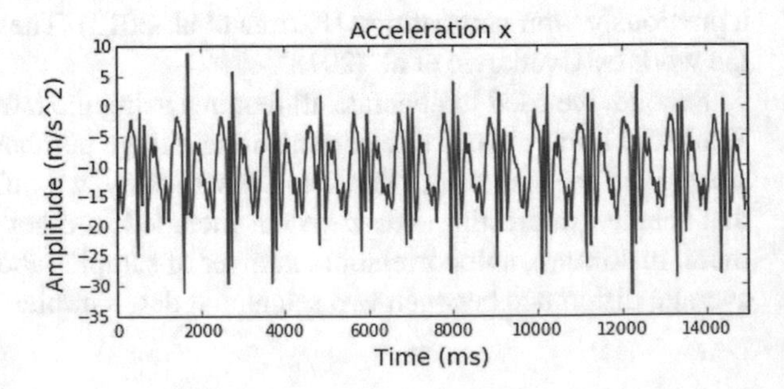

Fig. 15.1 Periodic pattern in a 15-s accelerometer segment during walking

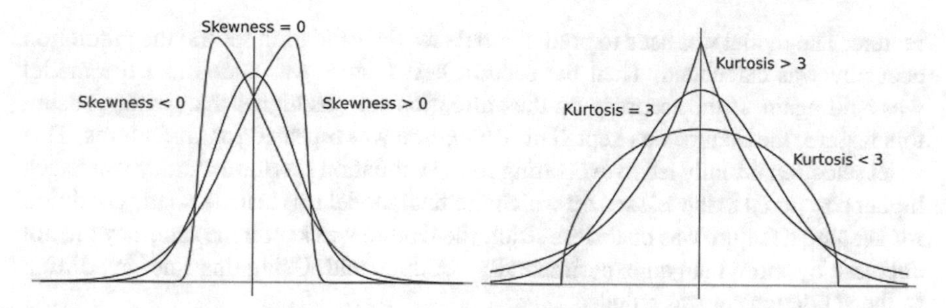

Fig. 15.2 Distributions with different skewness and kurtosis

- *Binned distribution.* A normalized histogram, which is essentially the distribution of the FFT magnitudes. First, the PSD is split into 10 equal-sized bins ranging from 0 to 25 Hz. Then, the fraction of magnitudes falling into each bin is calculated.
- *Skweness and kurtosis.* Calculated on the distribution-like PSD. Skewness and kurtosis describe the shape of the PSD. More precisely, skewness tells us about the symmetry of the distribution while kurtosis tells us about its flatness, as shown in Fig. 15.2.

  In total, 1696 features were computed and used in the subsequent steps.

15.4.3 Feature Selection

Since a relatively high number of features was computed, feature selection was used to remove redundant and noisy ones in order to reduce overfitting and speed up the training process. Our feature selection consisted of three steps.

In the first step, the mutual information between each feature and the label was estimated (Mutual info score 2019), where larger mutual information means higher dependency between the feature and the label.

After the features were sorted according to mutual information, correlated features were removed based on the Pearson correlation coefficient (Pearson correlation coefficient 2019). This showed that roughly half of the features are redundant, which was expected because different data streams contained similar information. Since calculating the correlation of all feature pairs was computationally too expensive, only 100 features were taken at a time, starting with those with the highest mutual information with the label. Correlation was then calculated for each of these pairs. If the correlation was higher than a certain threshold (experimentally determined as 0.8), the feature with the lower mutual information was discarded. After that, the next 100 features were added and the correlation between each pair was calculated again.

In the final step, features were selected using a greedy "wrapper" algorithm. A random forest (RF) model was first built on the train set using only the best scoring feature. The model was used to predict labels for the validation set and the prediction accuracy was calculated. Then the second-best feature was added and the model was built again. If the accuracy on the validation set was higher than without using this feature, the feature was kept. This procedure was repeated for all features. This strict selection initially led to overfitting to the validation set (the accuracy was much higher compared to the test set, on which the final model was tested), so the condition for keeping a feature was made less strict: the feature was kept if the accuracy did not decrease by more than an experimentally set threshold. Using this rule, overfitting to the validation set was reduced.

The best-performing features came roughly evenly distributed from each of the three categories, with those coming from the de-rotated magnetometer being the most significant, followed by those from accelerometer and gyroscope. The magnetometer came as a surprise, since it is the accelerometer that usually takes the spotlight in similar problems. It might be explained by the fact that the magnetometer is especially useful for transportation classification.

15.5 Machine Learning

In line with the principle of multiple knowledge mentioned in the introduction, we built an ensemble of ML models followed by a HMM (Fig. 15.3). The ensemble consists of ten base models: nine models that take features as inputs, and one deep neural network (DNN) that takes spectrograms as inputs. Of the nine feature-based models, one is also a DNN, while the other eight were trained with different classical ML algorithms. The output class probabilities from the base models are fed into a Meta model, which outputs a class prediction. Finally, the class prediction of the

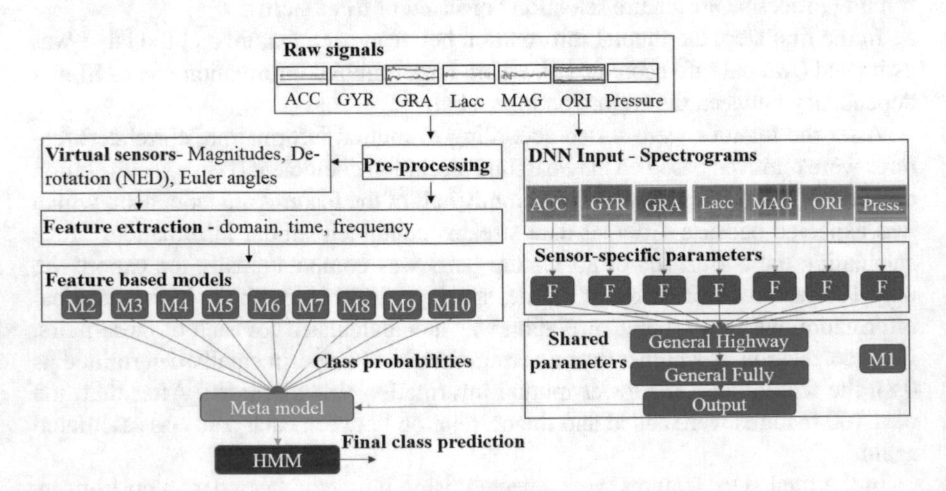

Fig. 15.3 Our machine-learning architecture

Meta model is corrected by the HMM, which smooths the predictions. The details of each step are presented in the following subsections.

15.5.1 Base Models of the Ensemble

The feature-based models are built using features extracted from the sensor data as described in the previous section. After the feature extraction, the data was fed to the following nine ML algorithms: DNN, Random Forest, Gradient Boosting, Extreme Gradient Boosting, SVM, AdaBoosting, KNN, Gaussian Naïve Byes and Decision Tree. The algorithms were used as implemented in the scikit-learn python library. The models' hyperparameters were tuned using twofold randomized parameter search. For the feature-based DNN, we experimented with different architectures, and the best performing was the architecture with 2 fully connected dense layers with 256 and 128 neurons.

For the spectrogram-based DNN, we again experimented with different architectures, including convolutional neural networks (CNNs) and long-short term memory networks (LSTMs). The final architecture is depicted in the right half of Fig. 15.3. For each category of raw sensor data, i.e., 3D acceleration, 3D gyroscopes, 3D linear acceleration, 4D orientation, 3D magnetometer and pressure data, a spectrogram representation is calculated using the Fourier transformation. The spectrograms are represented as vectors with dimensions P $\times$ T $\times$ N. P represents the number of spectral bands; T represents the time for which the spectral power is calculated; N represents the number of axes for the specific sensor type (e.g., the accelerometer has three axes, the orientation sensor has four axes and the pressure sensor has only one axis). The vectors are used as input to a fully connected DNN. The first layer of the network is a sensor-specific layer, which learns sensor-specific parameters. There are 128 neurons for each sensor type, thus there are in total 896 (7 $\times$ 128) neurons in the sensor-specific layer. The output of the sensor-specific layer is merged using a shared Highway layer, which is followed by a fully connected layer with 1024 neurons. The output of the model is obtained from the final layer with a softmax activation function yielding a class probability distribution.

To avoid overfitting, L2 regularization and dropout methods were used for all DNN models. The dropout probability was set to 0.3. The training was fully supervised, by back propagating the gradients through all layers. The parameters were optimized by minimizing the cross-entropy loss function using the Adam optimizer. The models were trained with a learning rate of 10^{-4}. The batch size was set to 2000.

15.5.2 Meta Model of the Ensemble

The Meta model takes as inputs the class probabilities output of each of the ten base models. We evaluated Meta models built with seven ML algorithms: Random Forest,

Gradient Boosting, SVM, AdaBoosting, KNN, Gaussian Naïve Byes and Decision Tree. Each Meta model was trained on the validation set and evaluated on the test set. The models' hyperparameters were tuned using twofold randomized parameter search on the models' training data. The best-performing model on the test data was picked as the final Meta model.

15.5.3 HMM Smoothing

Classification accuracy can be improved by considering the probability of a classified sequence. For example, the classified sequence: Train, Bus, Train, Train, Train, makes little practical sense, particularly in a short time interval, e.g., a couple of seconds. A misclassification of the "Bus" instance is much more probable than the user switching from a bus to a train and back in that time.

In order to find and correct this kind of mistakes, we employed the HMM method. This method assumes that there are some hidden internal states (in our case activities) that emit some signal at each time step (in our case classifications). The parameters of such a system are both the probabilities of transitions and emissions. Both can easily be inferred from the dataset—all we need are transition probabilities between each pair of activities and the confusion matrix of the model. Having the parameters of the system, a sequence of sequential classifications is given to the HMM method, which returns the most likely sequence of internal states—activities.

There are two possible scenarios where the HMM method can be used. In the first case (see Fig. 15.4) the whole classified sequence is known in advance. In this case, the HMM method can be used directly. In a real-life setting, this corresponds to reporting the classified activities to the user with a delay (as the HMM method uses instances classified after the current one). Different delays were tested to determine the relation between the sequence length and the method's usefulness.

In the second case (see Fig. 15.5), we have the entire history of classifications, but we cannot see the future ones. In a real-life setting, this corresponds to reporting activities to the user as they happen. This can be implemented by using the HMM to predict only the last element of the sequence, while iteratively lengthening it.

The HMM method requires instances to be ordered in the correct temporal sequence. This was not a problem for the internal test set, where ordering was pro-

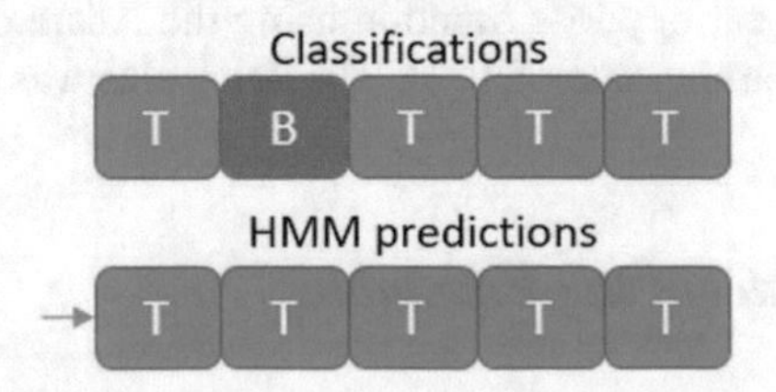

Fig. 15.4 Sequential classified windows, compared to HMM predictions. T = Train, B = Bus

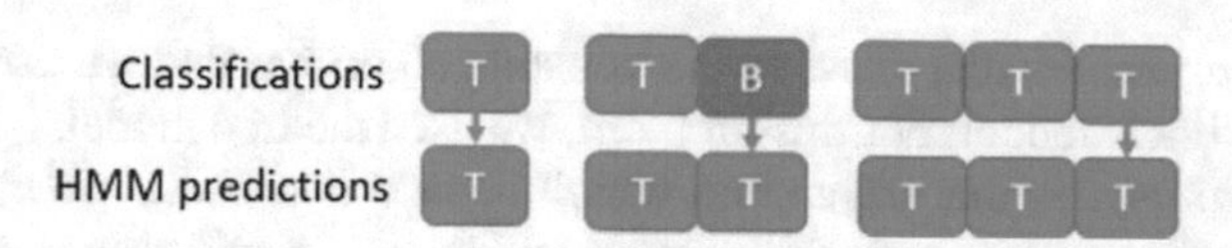

Fig. 15.5 Iterative HMM predictions. Sequence is iteratively lengthened as more of the history becomes known. Only the last instance in the sequence in changed after each step

vided. However, for the final test data the correct sequence of instances was unknown. To overcome this problem, we devised an algorithm that could automatically sort the 1-min intervals in the correct order. This was possible due to the fact that the last sensor reading of one window is very similar to the first sensor reading in the subsequent window, assuming the correct ordering. The algorithm iteratively searched for the next window that matched the above criteria. Doing so, it almost perfectly reconstructed the original ordering allowing us to use the HMM smoothing on the final prediction.

15.5.4 Single-Model Alternative

The previously described methodology—using an ensemble and HMM smoothing—is quite complex and thus risky. We could not ignore the possibility of a human error, and it was also possible that the labeled data available for internal testing prior to the submission of the results to the SHL Challenge was different from the competition evaluation data. Because of that we decided to submit two entries, one of which used a single model and no smoothing.

The single-model alternative used an Extreme Gradient Boosting model (XGB) (XGBoost 2018), which performed best of the base models in the ensemble, and is also often the best-performing algorithm in various ML competitions. XGB is an upgraded version of the gradient boosting algorithm. The implementation of XGB offers several advanced features for model tuning, computing environments (e.g., parallelization across several CPU cores, distributed computing for large models, cache optimization, etc.) and algorithm enhancement (e.g., handling missing values, optimal usage of memory resources, etc.). It is capable of performing the three main forms of gradient boosting (Gradient Boosting (GB), Stochastic GB and Regularized GB) and it is robust enough to support fine tuning and addition of regularization parameters. According to the author, the main difference is that XGB uses a more regularized model formalization to control overfitting, which gives it better performance.

In order to obtain best results, XGB requires careful hyper-parameter tuning. For comparison we first trained the XGB model with default parameter values and then improved the model through the tuning of its parameters. XGB has three major groups of parameters:

- *General parameters* that relate to the boosting algorithm that we use, commonly a tree or a linear model. For this problem, we used the tree model.
- *Booster parameters* that depend on which booster is chosen. We chose the tree booster, so the parameters in this section will be tree-specific parameters.
- *Learning task parameters* that decide on the learning scenario, for example, which evaluation metric should be optimized. Regression tasks may use different parameters than classification tasks.

Since XGB has more than 30 hyper-parameters, the parameter tuning process could not be done in one step, so we did it iteratively. This means that we optimized one or two parameters at a time while keeping the other parameters at default values. After finding the optimal value for the selected parameter(s), we fixed this value and started optimizing another parameter. For this iterative process, we started with the more important parameters as follows:

- We chose a relatively high learning rate. Learning rate defines the amount of "correction" we make at each step (each boosting round is correcting the errors of the previous round). So having a lower learning rate makes our model more robust to over-fitting and usually gets better results. But with a lower learning rate, we need more boosting rounds, which takes more time to train the model. Since we needed to fit a lot of parameters we started with a higher learning rate.
- We tuned tree-specific boosting parameters (max_depth, min_child_weight, gamma, subsample, colsample_bytree) for the chosen learning rate and number of trees.
- We tuned regularization parameters for boosting (lambda, alpha), which can help reduce model complexity and enhance performance.
- We lowered the learning rate to obtain the optimal parameter values.

We started by setting the initial values for the parameters that were going to be optimized. We chose:

- max_depth = 5
- min_child_weight = 1
- gamma = 0
- subsample, colsample_bytree = 0.8
- reg_alpha = 0.005

Note that these were just initial estimates and were tuned later. We took the default learning rate of 0.1 and checked for the optimal number of trees using the cv function of XGB which performs cross-validation at each boosting iteration and thus returns the optimal number of trees required. The obtained number of trees was 140.

The first parameters tuned were max_depth and min_child_weight as they were expected to have the highest impact on the model outcome. To start with, we set wider possible ranges and then we performed another iteration for smaller ranges around the best values obtained in the first step. The optimal obtained values were max_depth = 6 and min_child_weight = 1.

In the next step, we focused on tuning the gamma value. We used the parameters already tuned and searched for the optimal gamma value. The search showed that the initial value of gamma = 0 was the most suitable one.

Next, we tuned the parameters subsample and colsample_bytree. The optimal obtained values were subsample = 0.65 and colsample_bytree = 0.75.

Next, we applied regularization to reduce overfitting, even though the gamma parameter already substantially controls the model complexity. The best found value was reg_alpha = 0.0001.

Finally, we reduced the learning rate and increased the number of trees. We used learning_rate = 0.01 and number_of_trees = 5000. With this step, we added a significant boost in performance and the effect of parameter tuning became clearer. For comparison, the accuracy obtained with default parameter values was 88.3% and after parameter tuning we obtained the result of 90.2%.

15.6 Experimental Results

The first step in developing our approach was to decide on the temporal length of the windows to classify. Different window lengths were tested in order to determine the optimal one. Window length of 1 min was proven the best, as shown by the highest accuracy in Table 15.1. Only frequency-domain features were used in this experiment to train a RF model, as these were the fastest to compute and evaluate.

All subsequent results were obtained by using the 1-min-majority labels. Per-sample accuracy was also computed, but it was consistently around 0.5 percentage point lower, which is insignificant compared to the accuracy gains when using larger windows. This can be attributed to long-on-average activities, due to which activities very rarely changed in the middle of a given window.

Accuracies of the RF model used in feature selection are given in Table 15.2. One can note that the accuracy on the test set increases with each step. The accuracy on the test set is slightly higher than on the validation set in the first three lines, presumably because the test instances were easier to classify on average. In the last line ("Wrapper Strict"), the accuracy on the test set drops, suggesting that we overfit to the validation data. The difference between "Wrapper" and "Wrapper strict" is in the threshold to select a feature. The threshold was higher in "Wrapper Strict", so fewer features were selected—apparently such that did not generalize very well

Table 15.1 Accuracy using frequency-domain features computed on windows of length 5, 30 and 60 s. Random Forest was used to train and evaluate the model

Window length (s)	Accuracy (%)
5	67.1
30	72.5
60	74.9

Table 15.2 The number of features kept and accuracy of the RF model after each step of feature selection on both validation and test sets

	Features	Accuracy (%)	
		Validation	Test
All features	1696	78.9	82.2
Correlation removed	816	80.7	83.0
Wrapper	359	82.0	83.6
Wrapper (strict)	90	83.5	82.6

beyond the validation set. The same experiment was conducted with the XGB model, obtaining similar results: overall, the accuracy was higher, but the increase after each feature selection step was retained. The "Wrapper" feature set was thus chosen for the final model.

After choosing the "Wrapper" feature set, we proceeded to test different machine learning algorithms on the validation set. The resulting models were used as base models in our ensemble (with the exception of XGB, which was also used on its own). Table 15.3 summarizes the experimental results. The first column represents the accuracies of the 10 base models + the majority classifier. The middle column represents the accuracies of the complete ensemble using Meta models trained with different machine-learning algorithms. The rightmost column represents the accuracies of the ensemble with its predictions smoothed by the HMM method.

HMM-Past considers only the past data, HMM-2 and HMM-6 provide the output after a 2 or 6 time slots, while HMM-All had all the data as input.

For the base models it can be seen that the highest accuracy of 90.2% is achieved by XGB. This model was used for the SHL Challenge submission that placed second. The spectrogram-based DNN (DNN-Spec.) had 7.6 percentage points lower accuracy compared to the feature-based DNN (DNN-Feat.), suggesting that the selection of

Table 15.3 Accuracy on the internal test data

Test accuracy					
Base models		Meta models		HMM models	
Majority	16.0%	RF-Meta	92.0%	HMM-Past	94.0%
RF	84.8%	SVM-Meta	90.6%	HMM-2	95.0%
SVM	87.1%	**GB-Meta**	**92.2%**	HMM-6	95.5%
GB	89.5%	ADA-Meta	68.5%	**HMM-All**	**96.0%**
ADA	60.0%	KNN-Meta	90.8%		
KNN	81.5%	NB-Meta	85.9%		
NB	76.2%	DT-Meta	87.5%		
DT	74.1%				
XGB	**90.2%**				
DNN-Feat.	89.4%				
DNN-Spec.	81.8%				

	Still	Walk	Run	Bike	Car	Bus	Train	Subway
Still	95.3	1.1	0.0	0.2	0.3	0.3	2.0	0.8
Walk	0.3	98.7	0.2	0.6	0.0	0.2	0.0	0.0
Run	0.0	1.1	98.9	0.0	0.0	0.0	0.0	0.0
Bike	0.0	0.5	0.0	99.3	0.0	0.0	0.2	0.0
Car	0.5	0.2	0.0	0.0	99.4	0.0	0.0	0.0
Bus	0.9	0.2	0.2	0.0	0.0	98.8	0.0	0.0
Train	2.0	0.2	0.0	0.0	0.0	0.0	82.0	15.9
Subway	0.0	0.0	0.0	0.0	0.0	0.0	4.8	95.2

Fig. 15.6 Normalized confusion matrix on the internal test data for the model with highest accuracy, the HMM-All model

high-quality features importantly contributed to our success. For the ensemble using Meta models, it can be seen that the Meta model built with the Gradient Boosting algorithm (GB) had the highest accuracy of 92.2%.

Finally, the results using the HMM method show that the HMM significantly increases the accuracy up to 96%. When working with past data only, this benefit is halved, but it is still present. The small accuracy difference between HMM-6 and HMM-All indicates that a couple of time slots are sufficient to smooth the data. Also, HMM-Past has the lowest achieved accuracy compared to the other three HMM variations, indicating that classifying with some delay is better than classifying immediately. HMM-All was used for the SHL Challenge submission that placed first. Figure 15.6 presents the normalized confusion matrix for this model. The normalization is performed per row (i.e., the sum in each row is 100). From the confusion matrix it can be seen that the most problematic activities are Train and Subway.

For the SHL challenge submission, we trained the models on all the available labeled data and used them to evaluate the unlabeled competition evaluation data. To get an insight in the behavior of the "fully trained" model, we also trained an XGB model on 90% of the data and classified the remaining 10%, achieving the accuracy of 93.7%. We compared its classifications with the ones made with the model that was trained on only 50% of the data and found them very similar (they matched in 97.2% of cases).

Since the used dataset is large and the approach complex, it is worth mentioning the computational complexity. A workstation with a four-core 3.3 GHz CPU, 16 GiB of RAM and nVIDIA GeForce GTX1070 graphic card was used. The preprocessing and feature extraction required ~6 h, the model training required ~30 min, and the model testing required ~1 min on the internal test data (once the features were extracted). Additional 3 h were required to classify the competition evaluation data, mostly due to computationally expensive process of calculating the features.

15.7 Conclusion and Discussion

The SHL Dataset presents a uniquely large and sensor-rich dataset from real life. It provides an open platform for creating and testing various AR and other algorithms. By containing activities not common in AR, such as "Train" and "Subway", it opens new challenges for the AR community. The SHL Challenge was an effective way to jumpstart the research on the dataset, yielding the first solutions to the basic locomotion AR problem.

We approached the SHL Challenge systematically, first by paying attention to the organization of the work. The group of 11 people consisting of senior and junior researchers as well as a few of students was split into two teams. The first team—JSI-Classic—concentrated on processing the input data and applying one best classical ML method (Janko 2018), finally placing second in the competition. The second team—JSI-Deep—focused on deep learning, ensembles and smoothing, placing first.

At the beginning, the teams did not share information in order to encourage original thinking and maximize the benefits of multiple knowledge (Gams 2001). Only in the last weeks did they start exchanging ideas, data and software, so that both teams converged to the best solutions within their strategy. The strategy of JSI-Classic was to use a classical and safe approach, while the strategy of JSI-Deep was to use the maximally sophisticated and complex approach.

The high accuracy reported in this chapter can be attributed both to careful pre-processing, feature extraction and selection, as well as to successful use of multiple knowledge, implemented in the form of an ensemble, and on successful smoothing using the HMM method. If we assume 70% accuracy reported in the SHL Dataset authors' paper (Gjoreski et al. 2018) as the baseline, the contribution of preprocessing, feature extraction and selection, and selection of a good ML model amounts to roughly 20 percentage points, corresponding to the accuracy of around 90% achieved by the JSI-Classic team (Janko 2018). This can be broken down into 12 percentage points for the preprocessing and feature extraction, 1–2 percentage points for the feature selection, 5 percentage points for using XGB over the commonly used RF, and 2 percentage points for tuning its hyperparameters.

The additional 2 percentage points due to the ensemble and 4 percentage points due to HMM smoothing, corresponding to the accuracy of 96% achieved by the JSI-Deep team, do not seem much compared to the 20 percentage points of the JSI-Classic approach, but are still rather significant, having in mind that 100% is the absolute limit. The DNNs deserve special comment, since they are the name-sake of the JSI-Deep team and not commonly used for AR. The spectrogram-based model had a 7.6 percentage points lower accuracy compared to the feature-based model. This indicates that the features contain more information than spectrograms, which seems reasonable, since they contain additional specialized time-domain information, i.e., hand-crafted features that are based on the sensor's amplitudes. The spectrograms, on the other hand, only contain information about the change in the frequency bands over time. The DNN model using features was comparable to the best classical models, demonstrating that deep and classical approaches are comparable in this case.

The competition results were reported in average F-score, and we achieved the F-score of 0.94 which closely matched the F-score we achieved in our internal testing (0.96). Looking at the results of other competitors, one can see a large discrepancy between the results reported on their internal test sets and the results provided by the organizers on the competition evaluation dataset. The easiest explanation is that many competitors did not restore the labeled dataset to the original order, allowing them to heavily overfit to the training data as explained in the dataset description. Based on this, we can conclude that a key achievement for success was avoiding overfitting, although all the elements of our approach were needed to overcome the closest competitors.

Winning the first and second place suggests that the approach described in this chapter represents the state of the art in (locomotion) AR, both in technical and organizational terms. As such, it might be of use to researchers in the AR community.

References

Cvetković B et al (2017) Real-time physical activity and mental stress management with a wristband and a smartphone. In: Proceedings of the 2017 ACM international joint conference on pervasive and ubiquitous computing and proceedings of the 2017 ACM international symposium on wearable computers. ACM

Cvetković B, Janko V, Luštrek M (2015) Demo abstract: activity recognition and human energy expenditure estimation with a smartphone. In: 2015 IEEE international conference on pervasive computing and communication workshops (PerCom Workshops). IEEE, pp 193–195

Cvetković B, Szeklicki R, Janko V, Lutomski P, Luštrek M (2018) Real-time activity monitoring with a wristband and a smartphone. Inf Fusion 43:77–93

Gams M (2001) Weak intelligence: through the principle and paradox of multiple knowledge. Nova Science

Gjoreski H et al (2016) Comparing deep and classical machine learning methods for human activity recognition using wrist accelerometer. In: Proceedings of the IJCAI 2016 workshop on deep learning for artificial intelligence, vol 10, New York, NY, USA

Gjoreski M et al (2018) Applying multiple knowledge to Sussex-Huawei locomotion challenge. In: Adjunct proceedings of the 2018 ACM international joint conference and 2018 international symposium on pervasive and ubiquitous computing and wearable computers, Singapore, Singapore, 08–12 October 2018

Gjoreski H, Ciliberto M, Wang L, Morales FJO, Mekki S, Valentin Roggen S (2018) The University of Sussex-Huawei Locomotion and Transportation dataset for multimodal analytics with mobile devices. IEEE Access 6:42592–42604. https://doi.org/10.1109/access.2018.2858933

Hung WC et al (2014) Activity recognition with sensors on mobile devices. In: International conference on machine learning and cybernetics (ICMLC) 2014, vol 2. IEEE

Janko V et al (2018) A New frontier for activity recognition—the Sussex-Huawei Locomotion Challenge. In: Adjunct proceedings of the 2018 ACM international joint conference and 2018 international symposium on pervasive and ubiquitous computing and wearable computers, Singapore, Singapore, 08–12 October 2018

Janko V et al (2017) e-Gibalec: mobile application to monitor and encourage physical activity in schoolchildren. J Ambient Intell Smart Environ 9(5):595–609

Kozina S, Gjoreski H, Gams M, Luštrek M (2013) Efficient activity recognition and fall detection using accelerometers. In: Botía JA, Álvarez-García JA, Fujinami K, Barsocchi P, Riedel T (eds)

Evaluating AAL systems through competitive benchmarking. EvAAL 2013. Communications in computer and information science, vol 386. Springer, Berlin, Heidelberg
Mutual info score. http://scikit-learn.org/stable/modules/generated/sklearn.feature_selection.mutual_info_classif.html. Accessed 13 Feb 2019
Olsson LE, Gärling T, Ettema D et al (2013) Soc Indic Res 111:255–263. https://doi.org/10.1007/s11205-012-0003-2
Pearson correlation coefficient. http://en.wikipedia.org/wiki/Pearson_correlation_coefficient. Accessed 13 Feb 2019
Ravì D, Wong C, Lo BP, Yang G (2017) A deep learning approach to on-node sensor data analytics for mobile or wearable devices. IEEE J Biomed Health Inform 21:56–64
Reddy et al (2010) Using mobile phones to determine transportation modes. ACM Trans Sens Netw (TOSN) 6(2):13
Roggen D et al (2010) Collecting complex activity data sets in highly rich networked sensor environments. In: Seventh international conference on networked sensing systems (INSS'10), Kassel, Germany
Ronao CA, Cho S-B (2016) Human activity recognition with smartphone sensors using deep learning neural networks. Expert Syst Appl 59:235–244. https://doi.org/10.1016/j.eswa.2016.04.032
Su X, Tong H, Ji P (2014) Activity recognition with smartphone sensors. Tsinghua Sci Technol 19(3):235–249
Sussex-Huawei Locomotion Challenge database (2018). http://www.shl-dataset.org/activity-recognition-challenge
Teng H et al (2018) Chiron: translating nanopore raw signal directly into nucleotide sequence using deep learning. GigaScience 7(5). https://doi.org/10.1093/gigascience/giy037
Tsfresh (2018). http://tsfresh.readthedocs.io/en/latest/. Accessed 13 Feb 2019
Wang S, Chen C, Ma J (2010) Accelerometer based transportation mode recognition on mobile phones. In: Asia-Pacific conference on wearable computing systems, Shenzhen, 2010, pp 44–46. https://doi.org/10.1109/apwcs.2010.18
Wang L, Gjoreski H, Murao K, Okita T, Roggen D (2018) Summary of the Sussex-Huawei Locomotion-Transportation Recognition Challenge. In: Proceedings of the 6th international workshop on human activity sensing corpus and applications (HASCA2018), Singapore
XGBoost (2018). http://xgboost.readthedocs.io/en/latest/. Accessed 13 Feb 2019

Printed in the United States
By Bookmasters